Growing Up British

Book II of
The Eagle Must Fly
An Autobiographical Trilogy

Gisela Scofield

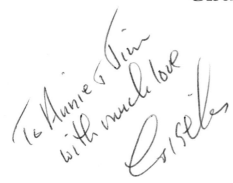

COVER PICTURE

A Scene in Chelsea Gardens
by Francis W. Moody 1824 – 1886

A picture dearly beloved by me and always hanging in our home. It was left to me in America in the will of Arthur Wheeler upon his death. A picture so purely and beautifully British that I felt it needed to stay in Britain and I made arrangements to allow that to happen.

Copyright © 2018 Gisela Scofield

All rights reserved.

ISBN: 1985204584
ISBN-13: 978-1985204584

DEDICATION

I would like to dedicate this book to my extended British family, an extension through love and not blood.

Especially to Dick and Betty Brewer who were always there as a friend to my Mother and to give help and advice to me when needed, throughout my growing up years. They were my rock and my stability for two months and beyond, when my Mother died in 1970. She was forty-seven years old and a widow. There is no way I could have cared for her at home in Wimbledon, with a new baby and a three year old in those last weeks of her life, without their constant help and attention.

CONTENTS

	Acknowledgements	i
	Preface	1
1	Rebirth	7
2	Six Months Later	15
3	Progress and a Strange Encounter	27
4	Moving On	35
5	A Miracle and a Secret	41
6	A Special Occasion, Growing Up, and A New Perspective	49
7	Christmas Preparations	57
8	Stille Nacht	65
9	The Promise	75
10	New Horizons	81
11	A Distraction	89
12	Growing Pains	99
13	An Alternate Option	105
14	Unfurling My Wings	111
15	Berlin 1962	121
16	Malta 1963	129
17	The Hands of God	139
18	Tripoli 1963	147
19	Fixing the Unfixable	159
20	Transitions	169
21	Flying on Wings of Memories	181

	Book III – Flying High	187
	Preface	189
1	New Beginnings	191

ACKNOWLEDGEMENTS

In truth, I have to admit to you my dear Reader, that putting together this Trilogy of my life would not have been possible without the love, encouragement, stamina, and incredible talent and dedication of my dear friend Janet Crane. She was there, steadfastly reassuring me at every turn and with every problem occurring for Book I – My Childhood in Hitler's Germany – and has never faltered for one minute with Book II – Growing Up British. She is a gifted writer and artist and I am forever in her debt for all she has done to bring this project to fruition.

Many thanks also to my friend Klaudia Thompson, who's cooking I miss dearly and who, like me, has roots deep in the Fatherland and who's last minute help with some finer points were invaluable in bringing this story to life.

*A Scene in Chelsea Gardens
by Francis W. Moody 1824 – 1886*

An English Rose

*How do you grow an English Rose
Sophisticate girl down to her toes?
You start with a child who's always been loved
Even when bombs rained down from above.*

*Transplant the child into a garden
Shower with love and watch her stalk harden.
A will of her own - her path is not yours
But she's always thankful you opened those doors!*

- Janet Crane

PREFACE

The heady smell of jet fuel and airport bustle came to a sudden end as the forward door to the Pan Am jet slammed shut and the handle slipped into the 'Locked' position. I was sitting in the first row of First Class with a new friend who, like me, was also following a dream to a new destiny. Her name was Chloe and she and I had both been accepted to start class, to train as stewardesses for Pan American World Airways. It was January 31, 1964.

We boarded the Pan American "Round The World Clipper Flight" together, in great excitement with four other trainee classmates from Sweden. The six of us stowed our luggage and snuggled gratefully into the comfortable softness of the spacious seats, carefully removing the sparkling white pillow and light blue blankets awaiting us in every seat. Chloe and I sat together in the first row behind the bulkhead with a large metal antique Clipper ship hanging from it.

She looked at me, grinned, and grabbing my hand said, "Can you believe we are really doing this and are on our way?" It was a surreal moment for both of us.

The reality of this giant step in our lives hit home with gusto. We were committed now with that slamming door, as the big Boeing 707 jet pushed back from its gate at London's Heathrow Airport en route to New York. Our eyes filled with tears as we held hands and left our loved ones and homeland behind. With just $200 in our pockets and hearts full of hope and anticipation, we gave in to trust for what lay ahead.

We met at gate 25 in Terminal 3 when we both checked in at the Pan Am counter. She had come from Devon with her parents to see her off and I had arrived with my Mother and Clifford Duits, a man I had known for years and dearly loved. To my delight, they had recently married and were there to see me off. We had special Crew Pass Tickets, and the Agent in the smart, soft light blue Pan Am uniform, smiled and introduced us as 'Class Mates' when she handed us our boarding passes. We would fly as "dead heading crew" in the First Class cabin all the way to New York. The agent grinned and had a twinkle in her eye as she handed us those first class boarding passes. "The back of the airplane is completely full and I have got to get you to New York today, so I have no alternative but to put you in first class – welcome to Pan Am."

We introduced ourselves to each other and to our excited but somewhat overwhelmed and slightly emotional parents. As my hand shook Chloe's and our eyes locked, I knew we would become close friends. That certain gut feeling does not happen often in one's lifetime, but when it does – you just know it is good and will last.

With the door now closed, the first class stewardess assigned to our cabin came around with glasses of champagne or mimosas, while the Purser welcomed us on board and went through the departure announcements. First in English and then in German because the flight had come from Frankfurt and many of the passengers on board had come from there. Then came the emergency announcement/demonstrations in English and German as the aircraft slowly made its way, in line, to the long departure runway on the far side of the airport perimeter.

Chloe and I locked arms in excited awe as the crew secured the cabin for take-off. We continued happily sipping the remainder of our mimosas out of tall stemmed glasses with the Pan Am logo on the side. Ten minutes later the Captain came on over the aircraft PA system, welcomed us on board and explained our route and how high we would be flying. He claimed the flight would be eight hours and twenty minutes in duration – a little longer than usual due to a strong head wind across the Atlantic. However, in spite of the head wind he anticipated a smooth crossing until we got to the Canadian coast at which time we could expect a slight "chop" of turbulent air. With that he wished us a pleasant flight - again welcomed us on board and directed the stewardesses to take their seats as we were the next in line for take-off. Our glasses were collected, hastily stowed, and the galley secured. We took in every move! Then the stewardesses took their seats at the exit doors and strapped themselves in. Minutes later the engines roared to life. When the captain released the brakes, the big airplane lurched eagerly forward. Then, after a long and ever quickening takeoff roll, lifted effortlessly into a wide open sky – heading towards the north and west.

I watched the London skyline gradually slip from view as the jet banked gently to the left. We had taken off towards the east into the wind, and our flight path took us over my Wimbledon home in southeast London, and on over the shining silver ribbon that was the River Thames. I eased myself back into the seat. The mimosa together with the pain medication I had taken earlier for my broken left wrist and badly bruised and painful left knee, had rendered me into a state of drowsy euphoria. All the excitement of the last hour or so had taken its final toll along with the mimosa. As a result, while the monotonous drone of the four big jet engines labored on and continued our lengthy climb into the heavens, I slipped gently into a deep sleep.

I did not hear the chime from the Flight deck alerting the cabin crew that we were passing smoothly through 10,000 feet on our way to the promised cruising altitude of 30,000 feet. It was now safe for them to get up and move around the cabin and start our elaborate meal service.

I was once again embarking on a monumental change in my life, and as I drifted deeper into the abyss of sleep, time seemed to spin backwards. I was back in another place and another time when I was seven years old and on board a life altering flight I had taken 17 years before. That amazing flight gave me an opportunity and opened a door that would enable me to embrace this one - and many more to come.

For a moment in my dreams, time was suspended and I was back there, as that small chartered British airplane took off from my still bruised and battered German homeland in 1948 and flew me to the stars......................

Preface

CHAPTER 1 –

REBIRTH

The monotonous drone of the airplane engines seemed endless – seemed an eternity. More than four hours had passed since we left my Homeland. I had exhausted all the ingenuity of a seven year old in flight. I had checked out all the seats, I had done tumble sets in the aisle, I had drawn pictures on every sick bag in the seat back pockets and looked out of every window at the endless sparkling ocean below. Now I needed a bathroom.

I roused my dozing Mother and informed her of this. She stretched and looked over at Captain Wheeler with a sleepy question mark on her face. There were no toilets on this airplane! He too, straitened up with a quizzical look at the cockpit and the relaxed Captain and First

Officer. They were lazily drinking the last of the coffee they had brought on board from the restaurant at the Hamburg terminal, and talking quietly. The curtain between them and the cabin had been left partially open, to encourage any interaction should it be needed. It seemed to be needed now. Captain Wheeler got up and made his way to the flight deck to get some input from the crew.

The Captain of the twin engine airplane turned, as he sensed the approach from the cabin.

"Captain Wheeler, Hello, Good afternoon, are you awake, how are you all doing back there?"

"We are all brilliant indeed Captain, but we do have a slight 'snafu.' How much longer before we reach Croydon do you think?"

"Let me take a look at the flight plan. Mmmmm, I'm thinking we should be on the ground in another hour and twenty five minutes. We will be crossing over the cliffs of Dover shortly, and then it will probably be another forty five minutes or so before we land. Is everything alright, you look a little worried?"

"All is well, except my daughter needs to 'see a man about a dog', and we were wondering how we could take care of that."

"Oh dear, that is a bit dodgy. The only thing you could do is have her use one of the sick bags in the back seat pockets – then just set the bag on the floor and we will take care of it upon landing."

My stepfather thanked the captain and returned to his seat and my worried Mother in whose lap I was now sitting. He explained to her the fix for my problem as she sat me up straight.

"Do you understand what you have to do Spätzchen? Go and get a sick bag out of that seat back pocket, and I will help you."

I looked at my Mother as though she had completely lost her mind. I jumped up and backed into the seat in front of her and on the other side of the aisle, grabbing my teddy and a pillow, and pressing them both tightly against my body. Then followed a complete meltdown.

"Nein Mutti, nein!" I shrilled in utter disbelief, glancing from my Mother to Captain Wheeler and back again as tears welled in my large horrified brown eyes and began to splash down my cheeks. Such shame – I could never perform such a thing in front of Captain Wheeler. And worse, what about those two strange men in the cockpit – I knew I would rather die first. I sat in my seat crying softly into my teddy, while my Mother tried everything to calm me down - to no avail. She held me and rocked me for what seemed like a lifetime, when suddenly a set of bells rang out in the cockpit. Soon after, the Captain turned around and lifting the curtain smiled as the aircraft started a steep descent.

"Ladies and Gentlemen, please fasten your seat belts we are starting our decent into Croydon Airport. We have picked up some tail winds as we crossed the cliffs of Dover and with their help we will be landing a little early – we may have some "chop" as we continue our decent as a result of those winds, but we should be on the ground in twenty minutes."

The aircraft hung as if suspended by invisible wires and then turned sideways fighting the wind. My Mother bit her lip nervously, readjusting her seat belt and held me tightly on her lap as the aircraft bucked and shuddered. Suddenly we hit the ground hard, became airborne and bounced again before the Captain gained control in the strong head wind. The aircraft bounced a couple of times more until the ground and gravity claimed it. The tail hit the grass and we had arrived. We were here in a world that was so new and so very different from anything I could have imagined or had known before.

Gaining control and momentum on the windy airfield, the Captain guided the aircraft swiftly to the small terminal building almost hidden completely from view by a large bank of River Birch bending in the wind.

As the aircraft came to rest, a small group of men in uniform materialized placing chocks around the wheels, and running a set of stairs towards the rear door. I do not remember if the aircraft steps had been completely pushed up to the door or not, before Captain Wheeler threw the lever and the door burst open. A sharp gust of wind catapulted my Mother out of the open door onto the old metal steps with me in tow. Swiftly regaining her composure she held tightly onto her hat with one hand and me with the other, then fighting the wind headed for the terminal building. The Captain of the aircraft joined my stepfather on the top of the stairs smiling and yelling at the ground crew to direct Mrs. Wheeler and her daughter to the Ladies Room, in the Terminal!

I have no recollection of the frantic dash from the airplane to the terminal – I vaguely remember my Mother saying "Toilette Bitte" to a young lady in Royal Air Force uniform sitting at a desk as we came in. She pointed to a door down the hall, and smiled knowingly, as we sped on.

There are no words in the dictionary to explain the mammoth relief I felt when we got there and I was sitting high on that throne. Now I had time to sit there and take in my surroundings. Amazing. How very clean. Unbelievable. My mind was racing, my brain in overload.

"Mutti, Mutti bitte komm mal rein. Ich muss dir was zeigen. Was ist Das?" I had slid off the seat and was putting my clothing back together. She slid the door open and looked in at me.

"What is what?" she asked smiling at my puzzled expression.

"Das" I said pointing at a shiny contraption attached to the wall. She came in and reached under the shiny top and pulled out a long roll of paper.

"This," she said smiling at my amazement "Is toilet paper, and you should have discovered it before you got dressed again."

I looked at the long paper runner coming out of the holder attached to the wall. Then I looked back at my Mother, my big brown eyes full of questioning excitement. I had never seen such a phenomenon.

"You mean, this is – no Mutti, you cannot mean it – this is TOILET PAPER and everyone gets some just like this in their booth. In EVERY booth?" I was jumping up and down in joyous but total disbelief. Then to my Mother's horror I started to unroll a big glob of the paper, tear it off and start sticking it up into my panties, under my skirt, and into my pockets.

"Was machst du denn?' said my Mother in disbelief. "What are you doing?"

"Mutti quick help me take some more, we will need some when we get to Daddy's house. He will need some too. They have so much of it here they will never miss it."

My Mother looked at me standing there stuffed like a mummy with brown toilet paper coming out of every crevice of my clothes, and burst out laughing. "Gisela this is England. Everyone has their own toilet paper here, you do not need to steal it."

"Are you sure?" I asked quietly as she began to unravel me from my burden. I hesitated, not sure if I could believe her or not.

"I'm very sure Spätzchen" she said, taking my hand and leading me back to the lobby.

"Oh Mutti," I said full of excitement now. "I think I am going to really like it here. Do you think that pretty lady that showed us the way to this toilet would let us live here with her, forever?"

My Mother was laughing happily as she led me back out to the lobby and my stepfather waiting there with our bags. He had a puzzled expression on his face as she tried to explain to him what all the excitement was about.

"Gisela feels that she does not need to go any further. She is content to live here in this terminal, she thinks this place is wonderful because it has toilet paper in every booth."

Captain Wheeler swung our bags onto a cart and threw my Mother a bewildered look "I don't understand – I have no idea what you are talking about Schatz" he said to my Mother, with a puzzled look at me. He then purposefully turned the cart towards the exit and the continuation of the first day of the rest of my life.

I did not know it but I was being reborn at age seven.

The Wheelers

Arthur's parents Walter and Ethel, My Mother Marina, and Arthur Wheeler (left to right)

1952

CHAPTER 2 -

SIX MONTHS LATER

"Mutti - Mutti wo bist du?" I called bursting into the back door of our house on Burbage Road, in great excitement. My Mother appeared from the direction of the dining room, her worried look turning into relief.

"Gisela where have you been? I have been worried sick. I called grandma to see if you were there but there was no reply. Look at your uniform – you look like you have been dragged through a hedge backwards. Bleib da in der Küche – stay in the kitchen with all that dirt and mud on you. Leave those muddy shoes outside and then take your dirty clothes off in the kitchen and wash your face and hands."

"Ja Mutti, aber ich muss dir 'was Unwarscheinliches erzählen - but I have to tell you something unbelievable." The words were pouring out like a river in full flood. "I have to tell you everything - where I was and what I was doing." I was turning my body into instant pretzel as I complied with her demands, wiggling out of my uniform and disposing it all over the kitchen floor. Our dachshund, Oskar was checking it out one piece at a time, tail wagging with such excitement at what he was smelling I thought for sure he was trying to wag it off. The adrenalin was on "high" and it was all I could do to check the excitement pouring out of me as big as Niagara Falls!

"Will you please slow down Gisela" my Mother said as she grabbed my arm coming out of the sleeve of my uniform shirt. "You are flinging mud all over the kitchen floor." But she paused, taking in my total outburst with sudden, intense interest.

"Mutti I fed and petted a horse today! Lots of them! You are not going to believe what happened on my way home from school. You see, when I got off the bus and turned into Turney Road, I saw the riding school coming down the road. They had been to Dulwich Park and they were heading home. That is when the most unbelievable thing happened – they turned into the Cricket Club gate right in front of me, next to grandma's house. Ausgerechnet da wo ich immer bin um nach Haus zu kommen. Right there where I turn to come home!" My eyes were on fire with excitement as my face surfaced after coming out of my uniform tunic.

My Mother had to smile in spite of herself at my

unbridled excitement. It did her heart good to see me so happy. Life had been somewhat of a challenge for me to say the least, since my arrival in London. I had been enrolled in a very special private school called Oakfield, quite close to my new home. The idea being to introduce me to children of my own age and hopefully, to learn English. The idea was intellectually sound, but had not taken into account the reality of the world after the war - at least the reality from a child's perspective.

The year was 1948 – three years since the defeat of Hitler and the end of World War II. However, to my fellow students - I WAS HITLER! Many of my peers had suffered mightily in the war, as had I – but now I was on THEIR turf, in THEIR homeland and I was the enemy that could not even speak English!

Life was hard. I had my hair pulled, my books and things thrown about. I was ridiculed and tripped up and kicked on the playground and in the halls just about every day. But I was tough and would not break. The staff and teachers did what they could to break up the fray and help to gently lead me into a safer space, but it was a very difficult time for me.

I shut down emotionally and retreated into a safe parallel world. I lost weight and was very pale, and my Mother and my new family were very worried about me. My sunny disposition and endless optimism were gone. They tried to think of anything and everything that might help me feel better and come out of my shell. My new stepfather, Arthur, even went to a breeder and purchased three dachshund puppies to add to our family. Max and Wolf went to live in Turney Road with my grandparents, and Oskar moved in with us. It did help. A lot. However, today was the first time the light really returned to my eyes. I was excited and almost normal again. My Mother's heart leapt at my sudden transformation. Dear God thank you, she thought. Somewhere a button had been pushed. She quickly pulled herself together.

"Mach schnell, go wash your hands and face and then come into the dining room," my Mother said grabbing my uniformed shirt away from the excited Oskar and picking the rest of my clothes off the floor. "Your tea is all set up and waiting for you. The baked beans are almost cold and the toast is hard, Du bist ja so spät. You are so late. Sit down and eat, you can tell me all about it then."

I did as I was told and promptly got the hiccups for talking with my mouth full. But the story continued to shoot out of me like a raft tumbling along the rapids on the Rio Grande. Just then the phone rang. It was my grandmother calling to tell my Mother more slowly and coherently what had transpired that afternoon.

My grandmother spoke slowly, so that my Mother could understand what she was saying. She explained that The South London School of Equitation had spent the afternoon in Dulwich Park and was heading home to Streatham Hill, when one of the eight horses went lame. It was still a long way home on hard roads, and they decided they needed to stop and call for help. They needed to call in to their office to ask that the trailer come to meet them and pick up the lame horse. They had decided they could not, should not, go on without help.

It was at this moment that I had arrived at the Cricket Club gate next to grandma's house. I turned in and stood in silent, total disbelief at the gate to grandma's garden that bordered that facility.

I LOVE horses. The universe stopped and I stood transfixed.

The group around the horses was totally engrossed in the problems at hand. I stood completely unseen and unnoticed for many minutes watching enthralled, as they checked and examined the pony's lame leg. Suddenly one of the horses started to wander in my direction, grazing unnoticed on the grass by the fence. As its rider became aware that her horse was moving out of range - my silent presence was discovered.

"Well Hello young lady, I did not see you standing there – Miss Peggy will not hurt you. Don't be afraid. Do you like horses?" The person on the other end of Peggy's reins asked smiling, while encouraging the pony to come even nearer to me. I was speechless and shaking – but in happy excitement, not in fear. Then she dug deep into the pocket of her jodhpurs and came up with two sugar cubes. "Here you go young lady – what is your name? Hold out your hand – very flat – and give these to Miss Peggy and you will be her best friend forever."

Without further ado, she grabbed my hand, flattened it out and put the sugar cubes on top, making sure my small thumb was out of the way before offering it to the pony. Miss Peggy must have sensed my fear, joy and excitement all wrapped into one, because she very softly nudged me in my middle as much as to say – don't be afraid I will not hurt you. Then very gently nuzzled my hand with her lips and took the sugar cubes off my open palm. The world spun on its axis – I was in heaven – I could not speak, but the happiness I felt was written all over me and my smiling face.

The leader of the group, Mrs. Parker, gently shook my hand and said, "Hello, do you have a name? If you do, Miss Peggy would like it very much if you would pet her right here on her neck, she really likes that as much as meeting new friends – so please tell her your name."

"Gisela!" I blurted out and gently touched the pony's soft neck.

"Eeesella?" That is an unusual name said Mrs. Parker smiling.

"Gi sel a," I repeated awkwardly, and lowered my hand to my side and my eyes to the floor at my feet. I was nervous now, because I could understand her, but was not sure that I could continue to do so and respond properly in English if she asked me too many other questions.

"OH - GISELA!" she repeated slowly, as a light suddenly came on in her head and she seemed to sense we had a bit of a language barrier. "That is a very pretty, unusual name," she said as she gently took my hand and started to turn Miss Peggy around and walk with both of us towards the waiting group.

"Gisela, my name is Mrs. Parker, come with me and

let me introduce you to my friends. We have a little problem with one of the horses – he has lost a shoe and he is lame" She straightened up and started to walk with an exaggerated limp to demonstrate the problem. "Do you understand what that is?" I looked at her with a puzzled frown and shook my head 'No', not understanding anything she just said. It did not seem to faze her at all as we approached the group who were now all watching us with great interest.

The next hour was pure heaven for me. I was introduced to all the riders and their horses one by one. I gave them each a little "bob" of a curtsy as I shook their hand – which is required by all good, polite little German girls upon introduction to an adult. As a result they smiled and invited me to give their ponies treats and lots of petting.

The group soon understood - that I did not! We indeed had a language barrier. I did manage to inform them that I belonged to the house behind the gate across from where they were settled, which instigated great excitement in all of them especially Mrs. Parker. I explained, the best I could, that my grandmother lived there, upon which Mrs. Parker handed Peggy's reins over to one of the other riders and asked me if she could please come with me to meet my grandmother.

As the two of us were walking up the garden path my grandmother appeared at the back door with two very excited dachshund puppies. She had heard a lot of commotion in the alley and came to check with the two little wiggly dogs in tow to see what was going on and if there was any sign of me. I was due home from school and she was getting worried. Mrs. Parker introduced herself and explained their predicament to my grandmother. She then asked if my grandmother had a phone, and if so could she please use it to call for help and transportation for the lame pony. I picked up little Max, my grandmother picked up little Wolf and we made our way with difficulty and two wiggling excited puppies, into the kitchen.

Fifteen minutes later we all returned to the group of horses and riders waiting patiently for news. Mrs. Parker explained that she had called the stable and they were sending Geoffrey, the stable handy man, to drive the trailer to meet them, pick up Mr. Pickle, the lame horse, and drive him home to Streatham Hill. I was allowed to mix and mingle with them and all the ponies for the next hour until Geoffrey arrived, when they loaded up Mr. Pickle and everyone started for home.

The task successfully completed, I watched them ride out of the gate onto Turney Road. They all turned and waved merrily as they started the long ride home to the stable on Streatham Hill. Grandma saw my little body deflate with sadness at their departure, and quickly fielded the tears by showing me what she had just pulled from the oven.

"Look what I had all ready for you for when you came home from school Gisela - your favorite shortbread! But now you have to really hurry home across the field, because Mummy will be very worried about you. I will telephone to let her know you are on your way and what has happened, but you will have to promise to run all the way."

I promised as I hugged her – kissed the dogs and grabbed the little bag of shortbread that she had all ready for me as I ran out of the door. It felt like I was flying in a sunny haze on wings of memory all the way home.

Three weeks later on a bright and sunny Saturday morning, I arrived at The South London School of Equitation, for my first riding lesson. I was dressed in new regulation jodhpurs, black riding boots, yellow polo sweater, black felt crash hat and red jacket. A bright red jacket was uniform for all members of the riding school under eighteen. The jacket would be uniform black after I turned eighteen. I felt as though I was living a dream. The very smell of horses put a big smile on my face – I was so happy.

Mrs. Parker welcomed me as did some of the other people I recognized from three weeks before. She took off my red jacket and said, "Let me show you around the stables and then we will find something for you to do to help us with the horses. We still have some chores to do before we can ride." I followed her around to one of the stables where Geoffrey, the stable hand, was grooming a pony.

"Geoffrey, this is Gisela and she is a new member of the Riding School. Could you please introduce her to Splash and show her what you are doing and then she can help you groom him. She will be riding him later this morning." Geoffrey smiled broadly and offered me his little finger to shake, indicating that his hand was dirty. The pony turned his head to see who had just interrupted the pleasure of his grooming session.

"Well right 'o then Missy, welcome to the Clan. Come on over here and let me give you this 'ere brush and you can brush from that side as I brush from this side. Would that work for you?" Geoffrey asked with a wide grin that creased his whole face. He smelled of horses, leather and hay and I liked him already.

"You will have to speak slowly and clearly Geoffrey – don't laps into any of that Scottish brogue of yours. Gisela is German and is only now learning English, so please make sure she understands what you are saying. I will be in the office Gisela, you can come and find me there when you are finished."

I nodded happily and followed Geoffrey around to the front of Splash to be formally introduced before I started my work. About an hour later I was able to lead the pony out of his stable to the hitching post, where I learned how to put the saddle on him and get him ready to ride. Putting the bridle on eluded me however, and I had to let Geoffrey accomplish that. Now I was ready to meet Mrs. Parker and retrieve my red jacket. She was ready too, and waiting for me.

"We are going into the big paddock this morning, Gisela. We are going to learn how to 'change the rein' and how to 'rise to the trot'. It is a little muddy in there today so I will have you on the leading rein and ride along with you. Splash is very sweet and should not give you any problem."

My heart leaped with excitement and happiness as Geoffrey helped me mount, tighten the girth, adjust the stirrups and settled me into the saddle. Then six of us students and Mrs. Parker rode around to the paddock where she put us through our paces. My heart was full, my life complete.

Later that afternoon I ran down Streatham Hill to the bus stop. There was no sign of the bus yet, so I went into the little Sweet Shop and bought a big bottle of Cream Soda. Then I sat on the window ledge of the shop's bow window and downed most of it, before the bus arrived. My world had returned to rotate normally – no longer spinning on its axis – I felt back in control of my life and my destiny. I smiled happily into the fizzy depth of bubbles as they hit my nose.

Gisela and Splash

Gisela in her Grandmother's Garden with Max, Wolf and Oskar

CHAPTER 3 -

PROGRESS AND A STRANGE ENCOUNTER

Since my introduction to the riding school and that world of horses and their associated people, my life took on a completely new perspective. I also had a new friend at school. Her name was Djagba Cudjo and she was black. Her parents were doctors and they were on an exchange program with the government of Ghana to a London hospital. Djagba was the only black child I had ever seen before in my whole universe, and because she was so different, she too was experiencing the same treatment from our peers as was I. This drew us together, and suddenly – I was no longer an outcast alone – I had a friend.

Djagba and I became inseparable. My parents were so happy to hear of her, that I was allowed to invite her to come and play with me at home on Saturdays if I did not go to the stable for a riding lesson. Then, sometimes I would go to her family apartment in Brixton to play with her there. English children from families like mine NEVER played in the street – you would play inside the house or the fenced in back garden. Never in the street! Her parents too were ecstatic that she had found a friend in her alien world that she liked and could relate to. It was very difficult at that time, for a black family to fit in. The only black people in most of post war Europe at that time were exchange students going to schools or universities to study. Small groups were dotted around in some of the larger cities in England, but none in my immediate area of south London.

I settled down, was more relaxed and happy, and as a result my English started to improve, a lot. My step dad had always insisted we speak English at home if possible. This would also help my Mother to learn the language. We could both understand most things, but we had difficulty with the spoken word. So now it was just when she and I were at home alone, that we spoke our native German. When we were all at home together as a family we would speak English and when we could not find the English word for something we would just say it in German! It was no problem because Arthur Wheeler spoke fluent German. It was because of this that he found himself stationed in British occupied Germany after the war. It was because of him that my Mother and I now found ourselves in this wonderful new life, in a new world without want or need. A parallel universe to the one I had known for the first seven years of my life.

It was thanks to him and this new life that I now went to school every morning like every other child my age. I was luckier than some – I went to a private school that my step father had picked out for me. It was better for me than the public schools he said, and more appropriate for a child from an upper middle class home, such as ours. It had smaller classes and he felt I would get more attention and help with learning "proper" English there. He just felt it would be a better start for me because he very badly wanted to get me into the very fancy private school that he had attended as a boy.

His school, Dulwich College was founded in the 1600s and had a sister school called James Allen's Girls School which was founded somewhat later in 1741. None of this however, was possible until I spoke perfect English and my grades were comparable with the high standards demanded in order to go there. I would be required to take an entry exam when the time came and pass with very high marks. Sometimes they would have 30 girls or more sitting the exam for maybe only two or three openings in a year.

For now I was a pupil at Oakfield School and I was dressed in a burgundy and gray uniform. It was Friday morning and I was standing at the bus stop waiting for the big red double decker bus to take me for the three stops to my school. It was a very brisk, chilly November morning, and I was bundled up in my gray school overcoat, felt hat with the Oakfield logo, and a two color woolen scarf wrapped twice around my neck. I was looking forward to the weekend because I was going to the stables first thing in the morning. I had progressed well with my riding lessons, and was now going to be allowed to join the riding school on their outings to Dulwich Park. I would be riding with Mrs. Parker who would have my pony on a leading rein. I would have complete control, but she would still have a hold on me should I need it. I was very excited, and stood alone at the bus stop in the cold morning air, day dreaming about what lay ahead the next day.

Croxted Road was long and only had light traffic on it as it stretched out in front of me while I awaited the number 3 bus that cold November morning. Suddenly I was startled out of my dream state by the presence of a man standing beside me. I had been vaguely aware of his approach but was too deep into my happy thoughts to pay it any attention. The man had been slowly walking towards me on the sidewalk, dressed in a long heavy overcoat and wooly hat. Now he stood beside me smiling and holding out his hand.

"Good Morning little girl. You are a pretty little thing. My name is Mr. Wood, do you have a name?"

I was flustered and jumped in shock as he stooped for my hand. "Don't be afraid, I won't hurt you. I was only going to show you a sweet little kitten that I have here in my pocket. You like kittens don't you?" he asked smiling a friendly, yet quite toothless smile. I hesitated, unsure if I understood him and how I should react. Suddenly, like a lightning strike, he grabbed my gloved hand in his and thrust it down deep inside his overcoat which flew open to expose his naked body and engorged manhood.

I was so startled that I was completely thrown off balance and grabbed at him as I stumbled for a hand hold in an effort to correct my fall and regain my feet. I ended up careening into the pole for the bus stop sign. As I wrapped my arm around it to stay on my feet I heard the bus approaching. I caught my breath as the big bus came to a gentle stop with the boarding platform right next to me. I had managed to regain my balance when the conductor leaned down and helped me climb on board. I had seen him before – this must be a regular route for him.

"'Ello Luv! Is everything alright – you look a tad tipsy. Did your Mum add a bit o' the old gin to your tea this mornin' then? 'Ere, sit right there by me and catch yer breath." And so the kindly old conductor with the heavy cockney accent took my three penny fare calming my nerves. I adjusted my hat that had been knocked askew during the ordeal in the reflection of the window, as we drove slowly past a man bundled up in a heavy overcoat and wooly hat walking down the street hunched over against the bitter cold and wind.

The conversation at supper that evening was light and uneventful when my Dad asked me how school was that day.

"Oh Daddy it was really nice – we went out in the garden and planted some flower seeds for Biology class. I'm very excited and can't wait for them to grow. Miss Lewis said Djagba and I have to check them every day to make sure they get water and the 'Eichhörnchen' did not dig them up and ruin everything".

"That will be a 'squirrel' Spätzchen – they are called squirrels in English, and yes you probably do have to watch out for them. It is winter, they are hungry and cold. Which reminds me, were you warm enough going to school this morning? I noticed you did not wear your boots like Mummy told you."

"Yes, I was OK." I said, suddenly remembering the incident at the bus stop. "But Daddy, I saw a man this morning at the bus stop, and I was so very sorry for him because he must have been really cold."

"Indeed," said Arthur Wheeler while absent mindedly concentrating on feeding his peas 'just so' up his fork. "What makes you think that?" he asked looking up at me while taking a long sip of his wine.

"Well, he was going to show me a little kitten, but I tripped when he took my hand and the wind blew his coat open and Daddy that man 'hatte gar nichts an'. He had no clothes on, and Daddy, I think that poor man 'war ganz verwachst' you know, deformed, and he never did have a chance to show me the kitty 'cos the bus came."

In my excitement to recount the memory I could not find the English words, explaining that I was so sorry for the poor man – not only did he not have any clothes, but to my seven year old innocence, he also was deformed. In my whole life I had never laid eyes on a naked man. The only two men in my memory that I had ever lived with, my Opa, Paul Sperling in Germany, and Arthur Wheeler, my new Dad, and neither one of them had ever presented themselves other than fully clothed in my company.

Suddenly my stepfather grabbed my arm with such force that my milk and my fork went flying across the table. His grip was like iron and I shrank in fear – not knowing what in the world I had just done.

My Mother jumped up in alarm trying to field the spilled milk with her napkin. "Mein Gott Arthur, what is she saying?"

I started to cry.

Arthur lifted me off my chair, took my hand and walked me gently to his big easy chair by the fire.

"Everything is OK Gisela don't cry, little one. I did not mean to scare you. Come here and sit by me and tell Mummy and me the whole story. Tell us everything that happened this morning at the bus stop. You did nothing wrong - just tell us everything that happened in German, so you don't forget anything."

So I did.

An hour later I had to repeat it all over again to a very tall, burly police constable who wrote everything down on a little pad. When he was finished we all accompanied him to the front door. I was holding my Dad's hand and pulled him down so that I could whisper something in his ear. The constable looked at my father who was smiling.

"Gisela wanted me to ask you to please be sure and give the man some clothes to wear when you find him, so that he would not be cold anymore."

"My dear young lady" the officer exclaimed laughing heartily. "You can be sure that I will find him, and when I do, I will take him to a place where they will give him some very special clothes of his very own to wear. They will even have his name and a number on them so that he and everyone else will know they are his." Then with a wink to my dad he retrieved his bicycle and wheeled it out of the gate, still laughing at his own joke.

He was still smiling on Monday morning, when to my utter amazement, I found him standing at the bus stop.

CHAPTER 4 -

MOVING ON

I had a wonderful weekend spending both days at the riding school. I got dressed early on Saturday morning in my riding school uniform, grabbed my riding Mac because it was cold and might rain, gave little Oskar a big hug and a treat and left the house. My parents were still asleep but knew my plans so I did not wake them. I walked about a mile and a half to Herne Hill where I got a number 68 bus up Norwood Road to Stretham Hill – then walked the half mile up the hill, to the riding school. It was quite a trek and usually took about an hour.

The stable was alive with activity when I arrived. The morning chores were in full swing and everyone was happy to see me. I was put to work immediately "mucking out" stalls which is hard work but I loved it. Loved the smell of manure and horses. Life was good. The association with horses, the stable and the harmony of everyone there, had made a big difference in my life and progress. Learning to ride and working with the horses had given me back my confidence and self-worth. It had been wonderful therapy.

My English was coming a lot easier now and I was starting to gain acceptance amongst my peers at school. The hazing and bullying stopped and I began to fit in. Because my language skills were improving, so was my class work. I was now able to complete my assignments and homework, and my grades went up. I was fascinated with geography and biology, and history blew my mind. I loved the outdoors and sports of all kinds and excelled at most.

I had been assigned a tutor and went to her house two times a week after school to work on math and science – both subjects gave me hives and I was doing poorly with them at school. I disliked both intensely and as a result they seemed to create a language barrier in my brain. Mrs. Hardcastle was very nice, patient and understanding, but I never learned to excel or like either subject. It was a necessity to nail them however, if I was ever going to be able to move up to James Allen's Girls School.

I went riding every spare minute I could and was getting quite good at it. I was now able to join the riding school in participating in horse shows and fairs or rides to Dulwich Park. We would also take a whole day away from the stable and go to Wimbledon Common. We would take a picnic lunch and spend a whole blissful day having steeplechase races around the very large open expanse of the common. The ride there and back could be quite traumatic. Mainly because it was very long, and took more than two hours to get there, in heavy traffic. It can be difficult to control the horses when you are riding two abreast in a group of eight or ten. Then you have to wait at a traffic light and find yourself and your pony, jammed between two double decker buses, sometimes in the rain. Being there was worth every minute of the stress of getting there however, and much fun was had by all, including the horses. Yes, Wimbledon Common was a lot more fun than Dulwich Park.

I loved to go riding in Dulwich Park too, and it was a lot closer and easier to get to. It was very scenic and picturesque riding through its beautiful old trees with their large canopies and heavy gnarled misshapen branches, some bending all the way down to the grass beneath. The green grass and colorful shrubs and flowers that were all so very well maintained were eye candy, but there was also another reason the park was so desirable. It had a very nice, sandy, riding trail all the way round it, which was easy on the horse's feet. You just were not able to deviate off the track to let the horses run - a slow canter was all that was allowed. I loved it so, at Wimbledon, when I could kick my horse into a gallop and sail along with him on the wide open spaces of the common, dodging trees and jumping bushes. That for me was heaven.

And so time passed and before I knew it, it was the end of another school year. I was suddenly a year older, and my dear friend Djagba Cudjo had to leave and return with her family to her native Africa. Her parents' assignment to King's College Hospital in London was over. It was very hard to say good bye. We had been through so much together – but we had both grown and learned a lot in so many different ways. We promised to stay in touch, and did for many years.

I had been given a bicycle for my last birthday and this opened up a whole new world for me. I was ten years old and my transition from damaged German refugee and war survivor, to normal English schoolgirl was now complete. My English was fluent. I had friends at school that I was now able to visit by riding my bike, instead of always having to take a bus or walk. The family had a car, of course, but my father picked his Dad, Walter Wheeler, up every morning and took it in to London to open the picture gallery of W. Wheeler and Son. The Jaguar would sit there in St. James's all day till they returned home to Turney Road at night. I was taught how to get where I wanted to go either by bus or train, or both. But now I was able to ride my bike to most places of interest to me.

This was very special. I loved riding my bike and would do it often, just by myself. I did not need to be with other kids – I was an only child and liked, and was happy, in my own company. I was lucky too, to live in a very pretty area of London south of the river Thames, called Dulwich. It was a tad off the beaten track and traffic was minimal. I would ride to the Village Tuck Shop and buy a piece of my favorite cake or candy. Then I would ride through the Toll Gate to Dulwich

Common and on, to my girlfriend Ashley's house. I loved visiting her on a weekend. She had two brothers and there was always something fun happening at her house. Her Mum loved to cook and bake. On cold wintery afternoons we would sit in their living room next to a roaring fire and eat crumpets and homemade jam with clotted cream. In the summer time I would ride home through Dulwich Park. Through the big beautiful old oak and chestnut trees to the grassy edge of the lake, sit and feed the ducks and pigeons. I was happy and care free.

It reminded me of another time, long before in another life time, that I would sit outside like this in beautiful surroundings and just enjoy nature. Then, I would sit very quietly in the long grasses at the edge of our field and wait for the Storks to come out to feed. We had two Stork families nesting in the barn roof and they were such fun birds to watch strutting around the tall grass on their long gangly legs, feeding on crickets and bugs of the fields. I never told anyone, but I was always hoping that one day I would catch one with a baby hanging, wrapped in a blanket from its beak! But I never did.

Now, here in Dulwich Park, I'd sit very quietly or lay back against an ancient oak tree. Just soaking up the smell of the soft newly mowed green grass, listening to the sounds of bees and birds and squirrels all busy doing their thing. I was so lucky.

I was safe.

I was never hungry.

I was always clean.

I was loved - I was happy - and time marched on.

Gisela and her new Dad Arthur Wheeler

CHAPTER 5 -

A MIRACLE AND A SECRET

The nightmares – those tricks of memory that often catapulted me back to my childhood in Hitler's Germany, slowly faded with time in my new homeland. The dark, traumatic days of death and survival that we had endured during that time no longer haunted me in my dreams. Now I had other challenges that arrived with lightning and nail biting speed.

In the spring of 1952 I sat the exam of my life.

However, the Stars were once again all aligned for Gisela "Spätzchen" Sperling – now Wheeler.

I sat the dreaded Entry Exam to get into James

Allen's Girls' School, in Dulwich - or JAGS, as it was affectionately known by all.

I NAILED IT!!

How that was possible, I have no idea! I was up half the night before studying - my brain in a lather.

My Mother finally came into my bedroom in the early hours of the test day, removed that sneaky little dachshund Oskar from under the covers, where he was fast asleep. She closed up the reference book that had fallen off the bed into the heap of others on the floor, and turned out the light. Then she carried little Oskar down to the kitchen and covered him up with a blanket in his own bed. She poured herself a glass of milk and carefully carried it up to the bedroom she shared with my dad. The stress and tension of the morrow also keeping her awake, way into the next hour.

When seated the following morning in that beautiful school, in a classroom next to the swimming pool, with twenty seven other hopefuls - I became painfully aware that I had studied all the wrong 'stuff'!

The school had sent a letter with instructions on where, and how, the exam would be conducted. The letter added – almost as an afterthought - that there were only two openings to the school this year. It ended with best wishes and good luck to Gisela Wheeler.

What to do! The only thing you can do, I told myself, as the pen slid out of my clammy, panic stricken fingers.

Your very best.

Take a deep breath.

You can do this

Don't freak out.

You have three hours!

SHEZZAAMM!!!!!!! It was truly a miracle.

Years later this result would control every future job interview and open the door to endless opportunity and possibilities that I could never have imagined.

Two months later I found myself in the school shop being fitted for my navy, red and white uniform. The main pieces, like the coat, tunic and shorts for sports were navy. The shirts were white and things like the ties and hat band had echoes of red as did the crest on the blazer pocket. The summer dress was blue and white check, and the gym knickers were black with elastic just above the knee. Perfect for hiding your hankies.

I felt very special. I was a JAGS Girl, and because of that – now, I really WAS special.

I rode my bike the three miles to school every day and home through the village via the Tuck Shoppe. I made new friends, learned how to play rounders and tennis in the summer and field hockey and netball in the winter. I was really good at anything to do with a ball - I had a very good eye, and played on all the school teams with enviable results.

At one rounders game I kept hitting the ball out of the field twenty-two times without them able to tag me out. The visiting team decided to "call" and sacrifice the match after only three quarters of the game was over. I was the last man standing and the score was thirty-four to two. They had a two hour bus and train ride home to North London and felt it was futile to go on. They were very gracious and thanked us for a good match and delicious tea, but warned that the next match we booked with them would have a different ending. We were all grinning with pride as we escorted them to the school gates and the bus stop.

I continued to excel in sports and became team captain of the middle school tennis and hockey teams. Academically, however, things were very different. My grades were mediocre and I did not work as hard on my classes as I did on sports. I joined the local tennis club which I loved because it was right across the street from my house on Burbage Road. The courts were grass of course, and red clay if it rained or for winter play, and I never had to wait too long before I was asked to join someone for a singles or doubles game. In the winter my dad and I played squash in the club house.

My dad's best friend, Dick Brewer, would come over once a week after work to play squash with him too. He had gone to Dulwich College with my dad when they were boys, and he was really good. He was four years younger than Arthur and had been best friends with Arthur's younger brother Reg who was in the Royal Air Force. Reg had died when his airplane crashed somewhere in the Irish Sea while on a training flight about one year before the war ended. His wife, now widowed, was my new Aunt Vera, and lived in Turney Road with my grandparents, after his death.

Dick's wife, Betty, would meet us at the house when we got home after a game. They would stay for dinner and then play cards, Samba Canasta - way into the evening. They were both really nice and I quickly learned to love them. I called them Auntie Betty and Uncle Dick. My Mother and Betty became very close friends. We spent a lot of time in each other's company and they became a big part of my life. They had a little girl called Carol who was really sweet and loved playing with Oskar. The fact that we became so very close over the years and Dick and Betty embraced my Mother and I as 'family', is quite amazing to me as I look back on it, because of the secret of which they were a part and we knew nothing.

Unbeknown to my Mother and me – the English extended Family Wheeler had a secret.

When Captain Arthur Edwin William Wheeler was deployed to Germany in 1945 he had a wife!

Her name was Mickie and she and Vera were close friends. They were part of a group of six friends that all grew up together. They met at school. Betty and Vera at their local private school of St Martin's in the Fields and the three guys Arthur, Arthur's brother Reg, and Dick all went to Dulwich College of God's Gift. Mickie moved into the neighborhood as a child and lived near Vera and they became friends. The six of them, Arthur, Reg, Dick, Mickie, Vera, and Betty all knew each other for years, growing up in the same area of South London around Dulwich and West Norwood. They later married and often did things together as a group – even going on holiday or spending weekends in the country. Then came the war and the boys were required to join the fight. Arthur went into the Army, Reg the Royal Air Force, and Dick went into the Navy.

I do not know to this day if my Mother was aware of Mickie. I suspect, knowing my Mother as well as I do, that she did not. I did not learn the secret of her existence myself until many years later, when I was in my mid-twenties.

It seems that Arthur had telephoned his parents in the New Year of 1946 and informed them that he had found the love of his life. Would they please, gently inform Mickie and start divorce proceedings. He would be home at the end of the year to sign the papers.

I can only imagine the heartbreak and confusion this must have caused Mickie. Especially, because she discovered that she was pregnant after Arthur had deployed to Germany.

It remains a mystery if Arthur ever knew this. Had he known - would it have changed anything. I doubt it. Poor Mickie's pain would continue because she lost the baby in her fourth month of pregnancy. Not however, because of the shock of Arthur's questionable behavior and deceit, but because the baby was deformed and not well - it was a blessing. She decided to leave London and return to her family in Oxford.

When Arthur returned to England and wired my Mother to gather her and my paperwork for our trip to a new life, all this misery and unpleasantness was behind him. Thanks to his loving - if totally devastated parents - and Dick and Betty, who helped handle everything with Mickie. The secret was never to be mentioned again.

Years later, when I talked to Betty about it she had an interesting perspective on the whole thing. In 1949 my Mother too, became pregnant and lost the baby early in the pregnancy. Betty thinks that there was something medically wrong with Arthur and he knew it. He loved and always wanted children, but was destined not to have any of his own. So when he met and fell in love with my Mother – I was the icing on the cake. He and I had become very close the last few months of his deployment to Hamburg, I too, was devastated when he left and was returned to England.

Betty had a good point, an interesting idea and it made perfect sense.

Thus totally unaware of all of this, for me, time marched serenely on.

A Miracle and A Secret

CHAPTER 6 -

A SPECIAL OCCASION, GROWING UP, AND A NEW PERSPECTIVE

About this time, a very big change was about to take place in Britain.

Our beloved King George VI died.

He had been sick for many months with lung cancer and blood clots in his legs. He had surgery and seemed to be rallying and was doing a lot better. However, not well enough to undertake a six month Royal tour of Canada, Australia and New Zealand that had been planned before his illness. Instead of canceling a multitude of plans and festivities overseas, it was arranged that his daughter, the Princess Elizabeth and her husband Prince Phillip, would take the long and stressful trip for him. It meant that they would have to leave their two children, Prince Charles who was three and Princess Anne who was not quite a year old, with Nannies and the other Royals for many weeks.

All the Royals were at the airport to see the couple off including His Majesty. The King seemed fine and although frail, was enjoying champagne and the festive occasion and seemed in good spirits. The Royal couple had elected to take a little 'side trip' as a mini vacation en route to New Zealand. The Princess had been given a small hand held movie camera for her birthday, and she was excited to try it out at the earliest opportunity. The atmosphere was lighthearted and festive as the special British Overseas Airways Royal flight became airborne and the Royals in the Rooftop VIP Suite toasted the departure with champagne.

The Royal couple stopped in Kenya, West Africa, where they had a vacation home in a Safari Park. Sagana Lodge had been given to them as a wedding present by the Kenyan Colonial Government. They had a great time relaxing and enjoying the animals – and the Princess played photographer. The highlight of the visit was to be a trip to the "Treetops Hotel" which was built into the boughs of an enormous fig tree overlooking a watering hole for elephants and lions.

When they arrived it seems the staff had prepared everything for the Royal visit – including special, extra toilet paper in the two private bathrooms. Sometime during the night a family of Baboons had managed to get into the "Tree House" and found the beautiful loo paper prepared for the Royal visit. Gleefully they hooked their arms through the rolls and amongst much screeching and excited chatter, they had a wonderful time throwing it at each other. The result was white loo paper draped exotically throughout the trees around the Tree House. A wonderful way to start their visit with much merriment and awesome pictures for future memories to come.

It was while here that they got word of the King's death. They returned home immediately. Elizabeth was very close to her father and this was devastating news. Unexpected too. She would not have agreed to go on this long a trip if she had any suspicion that her Father's health was not stable.

She left the country as a Princess and returned as a Queen.

This event affected my family greatly, because we were the only family in the neighborhood who had acquired a television set as a result. Princess Elizabeth would have her Coronation soon, and the BBC was determined it should be televised to the world.

The Wheelers were prepared.

The Royals, however, were not on board.

Even though the Royal Family was used to being in the public eye and on parade every time they went anywhere, they were not used to television cameras and reporters up so close and personal. They resented the intrusion. They felt it was an invasion of their private space and would not allow it to happen.

It took the BBC and Winston Churchill, who was our Prime Minister and beloved by His late Majesty, many months to convince the Royals this was a really good thing to do.

His Majesty had had a slow and painful start with Churchill. The man seemed to be hated by everybody in his own party, the right wing Conservative Torys. They were the blue bloods and upper crust of society in the late 1930's and early 1940's. When the opposing party of left wing Laborites insisted that Churchill was the only person trusted to lead them through a time of war and Hitler Hell, the entire country got behind him. By the time the war was over, the King got to know Churchill better – came to respect and trust him.

The Royal Family eventually gave in to the wishes of Churchill and the BBC to televise the coronation, but reluctantly and with much trepidation.

June 2nd, 1953 was a public Holiday and I think the whole world was in our house to experience it in black and white! We had at least thirty people in our house on Coronation Day; it was an amazing ceremony and a special, beautiful occasion. Elizabeth II was poised and regal and every inch a Queen.

I was growing up and started to notice a couple of things at about the same time.

Religion and Boys.

James Allen's was an all-girls school. Its brother schools, Dulwich College and Alleyns' School were all boys schools but were on different campuses, several miles apart. And never the three shall meet!

Except in church.

Until I got to JAGS I had never been introduced to a religion and I had never been baptized. In Germany we were too busy dodging bombs. According to my Swastika stamped birth certificate, my family was Lutheran but I did not ever remember anyone going to church. Probably because most of them had been bombed and lay in shatters for years as skeletons of their former beauty. Also, with parishioners widely scattered and without funds there was no inclination or ability, to rebuild. Any spare monies would be needed to build or repair more necessary things like hospitals and homes. Churches would have to wait.

Now, my days in school started with Assembly in the Great Hall. The school belonged to the Church of England and the Archbishop of Canterbury was its head, as was Her Majesty Queen Elizabeth II.

We would arrive at school in the mornings around seven thirty. We would then change our shoes and drop off our coats and hats in the cloak rooms in the basement. We each were assigned a hook above a bench with a cubbyhole under the bench for shoes. We were required to change into 'indoor' shoes before we were allowed to access the school corridors. Then we would hurry to our classrooms in silence to start the day. After roll call the bell would ring on the dot of eight o'clock and we would all file one class at a time, in silence and single file into the Great Hall.

The day started with a hymn, a reading from the Bible and prayers. Then the school program for the day was discussed and any information the students needed to know was given out. Awards, if any were presented, and collections for favorite charities' taken up. Then we filed out, one row at a time in silence, and dismissed to class.

A Chapel had been founded at the same time as the original boys school, Dulwich College in the 1600s. It was not to be a parish church, but an inclusion to the foundation of that school and the two to come, James Allen's Girls' School and Alleyns School. Many years later I would be baptized and confirmed there. Years later still, I would be married in that chapel after acquiring a special promissory grant from the Archbishop of Canterbury himself, needed, because of course, the little Chapel was not authorized to celebrate weddings.

I loved that little chapel. It was located in the fork of the road right at the end of Burbage Road and Village Way. There it was met by College Road and Gallery Road. It sat right in the middle, by the fountain. On Sunday mornings it had a full service with the College choir and all the boarders from all three schools present. This was a magnetic draw to view boys for one whole hour, secretly, quietly and unobserved. I did this and so did many of my friends. Then we would go home, giggle, and compare notes!

The only problem was that this service was so very crowded and I could not concentrate on anything other than the boys. I found myself very often totally distracted, resulting in the service itself and its message being lost. I liked the Vicar. He was very nice and would come every Friday morning to lead the Assembly at school and then teach an RT class - Religious Training. I had signed up for that. I was intrigued. One day he and I had a 'fireside chat' after class one day, at which time I confessed I had never been baptized. That idea took his breath away, but he was very sweet and non-judgmental when he learned of my background. He was silent for a while, deep in thought, and then he exclaimed with some animation,

"Gisela, do you realize that without having been baptized - I cannot even bury you!"

What an amazing, interesting idea. I was blown away. At age thirteen I had to think this information through, and maybe get some input from my parents. I was a heathen. I belonged to nothing.

It would be some time, however, before I would have reason to talk to them about it.

About this time I decided that every now and then, I would go to Eve'n Song at six o'clock on Sunday evening, instead of the crowded testosterone filled morning service. As a result I really enjoyed the simple, laid back service on a Sunday evening, and started to go regularly as it led up to Christmas. A very favorite time of the year.

A Special Occasion, Growing Up, and A New Perspective

CHAPTER 7 -

CHRISTMAS PREPARATIONS

Christmas at the Wheeler's was a very special affair.

My Mother loved Christmas, and being German it had to be celebrated on Christmas Eve. This was a totally new idea to my English family, but we worked the two different traditions into one unit and it was a success every year. The Burbage Road Wheeler's would have a large six foot Christmas tree in the living room with real candles in special little holders acquired from Germany. It would be lit on Christmas Eve for a Christmas party given by my parents every year. We would sing *Stille Nacht* and *O Tannenbaum* and we would have a wonderful night of feasting and festivities, exchanging gifts with friends.

The Turney Road Wheeler's would have a very large six foot papier mâché Christmas Cracker hanging up on the living room wall behind the couch. It would be filled with presents and cut down onto the couch after dinner on Christmas Day. Then I would get help from my Auntie Vera to pull it apart and to spill all of its treasures.

My Mother would ride the bus up to London every Friday to buy special things that she could not get in the Village or Herne Hill, our local shopping areas. She would go to Harrods in Knightsbridge, or Sainsbury, or Fortnum and Mason in Piccadilly. There she would find all the delicacies that she had been introduced to as a young adult in Germany, before the war. She craved and missed them, and would spend two months' worth of precious ration coupons on these treasures. Then she would come home on the bus, with her bags full of delicious Schinken and special German sausages and cheeses – including Limburger.

She would sometimes have to stand, for several stops, before she was able to get a seat on the crowded buses home during rush hour. Then she would get off the bus at Turney Road, walk to my grandparent's house and wait for my dad and grandfather to get home from the art gallery in St. James's, in the car. I would also meet them all there, because it was on my way home from school. They would sit and have a cocktail and talk about the day, politics and gossip and I would play with the dogs and drink Grenadine and soda. Then the three of us would pile into the Jaguar and my dad would drive us home to Burbage Road and the eagerly awaiting Oskar.

Sometimes my grandmother would marvel at how much my Mother was able to carry, and negotiate those crowded rush hour buses home, without incident. My Mother's eyes would light up with merriment, as she recalled the fact that the longer she was on the bus – the emptier the seats around her would become. Then she would go out into the kitchen and bring in one of her shopping bags – the one containing the cheeses, including Limburger. UGHA!

We all began to gag – even the dogs began to howl. No further explanation was necessary. That smell was disgusting and we could all envision my Mother sitting on a bus in London rush hour traffic, gloating, and able to spread out her shopping on empty seats all around her.

The first Sunday in December started Advent, and the first of the four Advent candles were lit on our Advent Wreath on the table in the dining room in Burbage Road. This was always the beginning of the Christmas preparations, along with the Glüwein, Stollen, Lebkuchen, Marzipan und Gemütlichkeit. This atmosphere was created by the German Christmas spirit - starting with the making of that amazing hot Glüwein. Once this was created upon the first days of December – we were in festive mood. Then the smell of amazing Stollen or Lebkuchen being baked was permeating the walls of Burbage Road – we were ready!! Christmas is here – well no not quite. We still needed Marzipan! It would come with the arrival of amazing, wonderful packages from Hamburg, sent with love from my Aunt Ilse.

Next came December 6 which was "Nikolaustag."

Christmas Preparations

It was the day Saint Nicholas would supposedly roam abroad in the German countryside and cities and place gifts of apples, oranges or candy in a shoe that good children had placed in a window the night before. This could be very scary for those children who had not been good, because in that case he would leave you a piece of coal. Nikolaustag was celebrated in many different ways depending on where you lived in Europe, but this is what we did in Wilhelmsburg.

I can attest to the fact that this can be quite emotionally traumatic. It happened to me December 6, 1944. We were still living in the cellar of the bombed out building in Wilhelmsburg. My family had nothing to put into the shoe – I did not even own a shoe!

I borrowed a shoe from my aunt because I did not have any of my own – I was still wearing my Klapperschue - a piece of wood tied to my feet with thongs – created for us kids during the war when we had no shoes.

We had nothing.

So it was decided I should get a piece of burned out coal – just as a warning to be good or the Weihnachtsmann, Santa Claus, might not come!

It was a totally moot point, because I did not get anything for Christmas anyway. Santa Claus was nowhere to be seen - and the bombs kept falling.

We had nothing to give, but that did not affect me as deeply as getting a piece of coal from Nikolaus.

On Christmas Eve we lit the stump of a candle that someone in the cellar had managed to steal, and sang *Stille Nacht* as the bombs rained down around us. We were alive and together, and that was all the Christmas present we needed.

My parents also had a special routine on a Friday night. When Arthur and my grandfather would arrive in Herne Hill on their drive home from the gallery, they would always stop at the flower shop. Both families would buy beautiful fresh cut flowers for the weekend and the following week. My Mother loved the graceful long stemmed Gladiolas, or the large 'dinner plate' Chrysanthemums or the colorful, fragrant, Freesias. As a result they became very good friends with Cyril and his wife, the owners of the flower shop. One Friday night Cyril asked Arthur if he thought I might like to come with him some morning to go shopping for flowers for his shop at the Covent Garden market. Covent Garden was the old flower and vegetable market in the center of London where he always bought his flowers. I could come with him and help him pick out a fine Christmas tree for us. I was thrilled at the idea of that, but not too thrilled at the idea of meeting him at 4 o'clock in the morning to do it!

Oh my word, what an eye opening experience that was! The amazing hustle and bustle. The noise, the singing, the music, the smells - the atmosphere of that giant old market. So busy and alive at that hour of the morning. I had to stay very close and hold on to Cyril's hand so as not to get separated or knocked down by running feet and wagons. As we flew along through the mayhem, we passed a stall with a rainbow of beautiful flowers and an old lady leaned down and thrust a bunch of violets into my arms.

"Mornin' Missy," she said and laughed as we rushed past her. I squealed in delight as they hit me in the chest, I grabbed them with my spare hand and we continued our flight through the crowd to where Cyril wanted to be. I was beside myself with delight at all the different types of flowers and the myriad of colors that the good Lord had created for our pleasure. It was truly magical and the aroma of fresh cut flowers engulfed your senses at every turn. Years later I would relive the memories of that morning, when I went with a friend to see the musical *My Fair Lady* at the Covent Garden Theater next door to the Market. It was with Julie Andrews and Rex Harrison. That, also, was magical.

Suddenly the colors vanished and were replaced by a sea of vibrant green. We had arrived at the Christmas tree section. The smell of pine took my breath away. For a delicious moment I was back in the deep, dark, silent forest in my old German homeland and the village of Neuhaus.

It took a long time before Cyril was happy with a tree of my choice. We walked up and down the long rows of trees but every time I would excitedly declare 'this to be the one' Cyril would nix it. Invariably they would have too much width here, or not enough height there, or they just were not good enough according to what Cyril had discussed with Arthur. Finally we came to a meeting of the minds. It was indeed special - seven feet tall and a perfect shape. We hauled it with great difficulty onto our cart and wheeled it to the truck. Then we went back and did Cyril's flower shopping, and because it was Christmas he needed some holly and mistletoe too. I was mesmerized, enthralled with it all. As a child I had no idea how much time, planning and work went into the fairytale magic of the end result that is Christmas Eve and Christmas Day. Even Boxing Day has its own magic – the day after Christmas – and has nothing to do with any person doing anything physical!

This experience at Covent Garden would always be remembered by me over the years, as the perfect way to start the excitement of the Christmas season.

Christmas Preparations

CHAPTER 8 -

STILLE NACHT

I learned over supper that this Christmas was going to be different.

We were going to spend Christmas in Germany this year. In snowy Bavaria, at a tiny village called Grainau nestled cozily right up against the forest wall of the German Alps. It snuggles into the granite wall at the base of the mountain called 'die Zugspitze' - the highest mountain peak in Germany. The nearest town is Garmischpartenkirchen south of Munich. The countryside is breathtaking in the summer time with its tall peaks and thick forest, but in the winter it takes on a special magic of its own. In the winter it wears a sparkling blanket of shimmering silver white crystals, the beauty of which can take your breath away.

I was beside myself with excitement. This would be the first time that I had seen real beautiful clean white, thick snow, since my exile from war torn Hamburg to the countryside of eastern Germany in 1944 - 1945. Sometimes the snow would be up to my neck there, in Neuhaus, which would not be so good, and I would not be able to go out of the house for days. But other days would be wonderful as I would join my village friends sledding and skating and generally frolicking in the pristine white stuff.

Christmas in the mountains of Bavaria - the majesty of the snow covered peaks, the silent sparkling silver of the tall pines with a tinkling, crystal clear bubbling brook at their feet, was truly magic. This would be the stage for our Christmas this year. I was beyond happy.

In London we got ready by packing winter woollies, boots, jackets and Alpine hats. Oskar went to stay with my grandparents and his brother in Turney Road. Little Wolf had died of distemper the year before, so it was just him and little Max now. Then we set out in the Jaguar for Dover.

We left London and found we were going to be somewhat early for the ferry to Calais, France. Not wanting to hang around Dover in the rain, we decided to make a side trip to Canterbury. The Cathedral is a sight to behold at any time of the year but at Christmas it leaves you spellbound. I had been learning about the Archbishop Thomas á Becket in history lessons at school.

Thomas á Becket was the Archbishop of Canterbury in 1162 when England was Catholic and controlled by the Church of Rome. He was having serious conflict with his king - Henry II, who wanted more 'say so' in the church and its Catholic, financial business. Henry became so frustrated with his stubborn Chief Cleric, that he let it be known amongst his Knights, that he would like very much to be rid of him. As a result, four of them rushed off to do the dreaded deed, confronting the Archbishop in his Cathedral, and then, near the High Alter, stabbed him to death with their swords.

Thus continued my history lesson in living color at ground zero – the Cathedral at Canterbury.

Sometime later, we proceeded on to Dover where we drove the car right onto the large car ferry. Then, because it was cold and raining, we went straight to the ship's restaurant to ride out the hour and a half ferry ride to Calais, in warmth and comfort. The twenty-five miles of the Straights of Dover which separate England from France, is no picnic in good weather at the best of times. It is truly vicious at this time of the year. The ship bucked like an angry steed as it was tossed by wind and rain and then dove deep into the troughs of wave after wave. As a result, many a poor passenger spent the trip prone on the pitching deck or hanging over the side desperately trying to hold onto their last meal without success.

Upon arrival in Calais, the weather continued to be cold and blustery with bouts of heavy rain. The ride through France, Belgium and Luxemburg into Germany was long, boring and uneventful. In the summer we would take the ferry to Ostend further up the coast from France - to Belgium which was a much longer ride at sea - about three plus hours. Then we would take the main road to Brussels and on to Aachen and the Rhineland or to the Mosel, a tributary of the Rhine. We loved to spend our summer holidays in the Black Forest and that part of Germany. Sometimes we would go to Hamburg and visit relatives and then drive through the beautiful Lüneburg Heath and on to the Hartz mountains towards the eastern side of the country. We always drove. Air travel was not yet as popular or as convenient as it is today. We would go every year and had done so since 1948.

Now it was winter time and we were on our way south to Bavaria as quickly as possible in the typical inclement weather for this time of year. We would stop to eat in tiny villages along the way. The routine was always the same. My Mother would go in and ask to use the rest room. If it was neat and clean we would park the Jaguar, get out and eat there – if not we would move on to the next place. My Mother was of the opinion that if the rest room was clean, so was the kitchen! Sometimes it took two or three stops before she was satisfied going through France and Belgium. She was not a fan of either of these countries, a prejudice she had carried from her German roots. We always drove through the countryside, never on the main highways or Autobahn. As a result it always took longer to get where we were going, but it was a more scenic drive. It would usually take two days, sometimes three and Arthur always did the driving. My Mother had never learned to drive the Jaguar, a 1946 model, which had a little idiosyncrasy. You had to double declutch every time you changed gears which could be quite tiresome and needed an experienced driver to handle it smoothly.

As we continued south, the rain started to turn to sleet and finally snow as we approached Frankfurt. The mood inside the car was also becoming more festive as we started to get into the Christmas spirit by singing Christmas Carols – some English and some German. It was dark and getting quite hard to see through the driving snow when we arrived in Garmischpartenkirchen.

We had stopped at a bed and breakfast hotel in a little town just inside the Belgium/German border on the first night of our holiday. The weather with its blowing snow had made the second day long and tiresome. It was difficult to negotiate the Alpine roads in weather such as this, in a car built to be driven as the English did, on the left hand side of the road. So we were relieved and happy to pull into an interesting looking Inn on the south side of town called Pension zur Post. It was beautifully decorated for the season. The entire staff was family, dressed in Dirdls and Lederhosen and was welcoming and friendly.

The lobby was warm and smelled of Christmas Stollen and cinnamon apples. An enormous Irish wolfhound was curled up snugly in front of a blazing fire in a large open fireplace. It felt good. It felt familiar – like home. I immediately went over to the hearth and introduced myself to the dog, who lifted his shaggy head lazily and smiled a welcome with his eyes and tail. His collar stated that his name was Prinzi.

The next morning we took our time over German

coffee and a continental breakfast of soft boiled eggs, Brötchen, Schinken and many different types of cheeses and German sausage. Prinzi, the gigantic beautifully ugly Irish wolfhound, sailed majestically back and forth with wagging tail and large wistful brown eyes, hoping for a treat. He was not disappointed for I could not resist slyly slipping him a piece of my sausage as he passed my chair.

After breakfast Prinzi gleefully rush out with us into the soft billowing snow, with the high mountains all around us, to find the car which was now snow covered and in a deep drift.

The owner of the Pension and his son were happy to help Arthur dig the Jaguar out of the snow drift to the accompaniment of excited barking, prancing, and flying snow. Thus, with the loud jubilant encouragement of a large shaggy dog, the job was accomplished in record time. As a result three men and a dog were soon warm and drying off by a blazing fire, and hot Glüwein was passed around with much merriment and Gemütlichkeit.

We left the happy Pension zur Post reluctantly around late morning with a promise to return another time, and drove off deeper into the mountains. We stopped for a leisurely lunch at a tiny wayside restaurant next to a shop selling Holtzschnitzerei along the snowy roadway. The delicate wood carvings were exquisite. We took some time talking, and marveling at the beautiful handy work of a third generation craftsman and his son. My Mother was enthralled and promised to return on our way home to purchase some of those delicate treasures - but now we must press on.

Sometime later we arrived in the sleepy little village of Grainau. We drove past the Mill by the stream, the little railway station with its lighted wooden clock tower and on to the blacksmith shop. Then the road took a sharp right hand turn, past the little village church with its onion top steeple, typical of Bavaria and this area of Tirol. Soon the climb became steep, uphill, and into the forest. We arrived at our hotel in the late afternoon. All was silent and still in the snowy gathering darkness, but music and soft lights glowed a welcome as we entered the wide Hotel Edelweis parking lot.

The Hotel Edelweis fit in perfectly to its snowy surroundings. It was a large wooden log house. A sharp pointed snow covered roof came all the way down to just above the first floor windows. All the windows from the street to under the eaves of the roof on the third floor had window boxes filled with evergreens, pine cones, red berried holly and twinkling lights. The entire front wall surface was painted with an Alpine hunting scene. It was truly picturesque and very Bavarian. The inside was just as pretty.

The lobby was warm and cozy with large, heavy, very old handmade furniture in the style of the mountains of Tirol. The furniture was heavily carved and was beautifully stained or painted in soft, gentle earth colors. The decor also had a lot of things made out of antlers and there was a large boar's head over the reception area counter. The walls had mounted deer heads sporting large racks of antlers. A ginormous softly glowing chandelier made of antlers hung like a halo from a heavy chain in the middle of the lobby. Old guns, knives, swords and helmets were hanging everywhere amongst other old metal antique artifacts. The old metal gave the walls a warm metallic glow as it reflected the gentle light of candles, the chandelier and an open fire.

Towards the back in the center of the lobby room was a very large fireplace made of ornate and colorful tiles. A large gnarled piece of wood was burning merrily and putting out immense welcoming heat. The fireplace had cozy little tables, couches and chairs on each side of it, close to the tiled walls. This was so that those sitting there could touch them and feel the comforting, radiating warmth after coming in from the cold, to enjoy some refreshment. While sitting there they could also enjoy a tall beautifully decorated Christmas tree full of handmade wooden ornaments and with tinsel hanging in glistening silver splendor from every needle. The unlit candles on each branch tip, hanging with eager anticipation awaiting their lighting on Christmas Eve, to bathe that beautiful tree in a halo of true warm splendor. The warmth of the fire brought out the aroma of the deep, dark woods of Bavaria.

To the left of the fireplace was a long, heavy wooden harvest table. It was the "Stammtisch" and was alive with noise and laughter – even music. This was the open, communal table, where all the local gentry would come at the end of the day to share stories, tell lies and brag about their prowess at hunting or fishing or anything else that came to mind. This merriment would take place accompanied by large Steins of local beer consumed in a haze of sweet smelling pipe smoke curling lazily towards the rafters. In a far corner between the bar and this lively group was an old man in leather britches. He had colorful braces holding up his lederhosen and a Tirolean felt hat with a large feather, strumming a wild German jig on an old violin. A friend in similar dress was accompanying him on an accordion. Every now and then the group would break into rousing, happy song.

I stood transfixed - inhaling the mood, the smells and the atmosphere - taking in every detail of the camaraderie, pure joy and goodwill. A giant old Grandfather clock, hidden from my view around the corner of the front desk, came to life and started to play a Glockenspiel and then boomed out the fact that it was five o'clock.

Arthur and my Mother were at the desk checking into the hotel, when an old man with a long shaggy beard came bursting into the lobby. He was dressed in green and gray leather hunting gear and wearing a felt hat with a large fluffy feather typical of the local Bavarian 'Trachten' uniform of his friends and peers at the Stammtisch. He made a very colorful spectacle as he burst though the heavily carved wooden door in a shower of blowing snow proclaiming a cheerful "Grüss Gott" to the world in general. He had a shotgun and a rabbit slung over his shoulder and a little dachshund at his heels. He shook the snow off his hat and shoulders and stamped his snow covered boots on the thick door mat; hung the shotgun on the special rack to the left of the door made for just such an event and carefully tied the rabbit below it. Then with a joyful gait of anticipation and a cheerful wave of the hand, he walked purposefully over to his friends who gave him a rousing welcome and handed him a large Beerstein.

The little dog followed his master for a few steps, tail

waging a greeting of his own and then deviated, making a beeline for the open warmth of the hearth. Hearing his master's happy welcome and conversation, he considered his job done for the day. He laid down, staring happily into the open fire and letting its warmth slowly melt the ice crystals coating his chest and ears.

"Grüss Gott." The Wheelers had arrived in the heart of southern Germany.

Stille Nacht, Heilige Nacht will unfold in due time so I happily joined the dog by the fire.

CHAPTER 9 -

THE PROMISE

My sixteenth birthday in February of 1957 dawned in the middle of a blowing snowstorm. The UK was in the throes of a winter the likes of which it had not seen in many a year. It held us in an Arctic grip. Heavy snow had fallen for some days – weeks really, on and off. Nothing mechanical moved. The city was paralyzed, immobile. Public transportation, which most people relied on to get to work or school, was holed up in their respective depots. Only the main roads were plowed but even they had surfaces of solid ice.

It was a Saturday and when I opened my bedroom

door to go to the bathroom to brush my teeth I could smell Kippers cooking in the kitchen. My Mother was already up cooking my favorite breakfast. I dressed hurriedly and ran down to that welcoming aroma and the smiling face of my Mother and the excited wagging tail of little Oskar.

"Happy Birthday Spätzchen," she said to me in German, and gave me a big hug. "Breakfast is on the table. Please go to the garage and get your father to come in and eat. That car is not going to start without a lot more warmth and encouragement. He has been out there for more than an hour without any luck and on an empty stomach." She and I still spoke German when we were alone, but it was becoming less and less frequent. I put on boots and my heavy winter coat and fur hat and opened the kitchen door. The wind burst in and took my breath away. Oskar barked and ran and hid in his bed under the sink. Some underwear that Mother had washed and was hanging on a clothes rack in front of the stove flew back into the kitchen and landed in the middle of the kitchen table on the bread and eggs. I slammed the door behind me, bit my lip and braved the elements.

The last few weeks had been difficult. It would take Arthur many minutes in the mornings to coax the Jaguar to life. Every night he would drive the car down the side alley to the garage at the bottom of our garden. He would back it in and then cover the engine with a heavy blanket and place a small gas heater on the sandy floor underneath it. The garage was an old poorly built wooden structure with no insulation to protect the car from the elements. It just acted as a cover from the blowing snow and could be locked against theft.

I walked the length of the garden through the thick snow in Arthur's boot prints, all the way to the gate and on to the garage. He was sitting dejectedly in the open door of the Jaguar playing with the starter. He had left the little heater under the engine overnight in the hopes that he could get a spark of life this morning - to no avail. The flame had blown out sometime during the night when the wind whistled and crept through the old wooden sides of the building. I smiled and went to hug him – he looked so dejected.

"Mummy sent me to get you in – breakfast is ready and it smells so good. She made Kippers for my birthday." He brightened, hugged me and wished me a Happy Birthday, slammed the car door and followed me out into the blowing snow. I helped him close the garage doors and throw the heavy lock. Then we retraced our steps through the long deep expanse of snow that covered the garden to the house.

My Mother had set a really festive breakfast table with flowers, cards and a big box sitting on my chair. New black riding boots! I had a big horse show coming up in the Spring down in Somerset, that I was really looking forward to. I was thrilled. I needed new boots.

The phone rang. It was my friend Jillian Clack wishing me a happy birthday and wanting to know if I felt like going to the pictures. A movie might be a good idea on such a cold, dark miserable day – but, how to get to the movie theater. There were no buses moving and the Jaguar also was being unreliable, otherwise Arthur could have dropped us off.

"We can bundle up and walk," Jillian said joyfully. "It will be fun! I'll phone a few other people we can have a party. Go out to The Taj for curry afterwards and celebrate your birthday."

The movie started at four o'clock. Four friends had managed to make their way to my house and my Mother had warmed everyone up with hot chocolate or Ovaltine. And so it was about two o'clock when we all trudged giggling and in high spirits into a wilderness of snow. We made our way into the middle of Burbage Road and turned towards the Village and the long trek to the theater.

When we got to Dulwich Village we were surprised to find quite a few people out braving the elements and doing their shopping - totally taking the lack of transportation in stride. We turned onto the High Street and three of us promptly fell in a heap as we crossed the curb. The High Street to Peckham had become a sheet of black ice overnight but we were in such high spirits and not paying attention - so that fact had eluded us. The fall just made everything more fun as we thrashed about and laughingly helped each other up.

We arrived at the movie theater at three forty-five and joined the cue of other brave souls to buy our tickets and get in. The movie theater was cold – no heat, but a young lady usherette welcomed us with her flashlight and a warm smile and showed us to our seats.

The movie was Elvis Presley in Blue Hawaii.

The seats were hard and the leather cold to the touch but we made ourselves comfortable as the movie started. Suddenly the Star appeared and I grabbed Jillian's hand as Elvis filled the screen and started to sing. My heart plummeted to my stomach and I froze with glee. We were transfixed. We all sat enchanted in that freezing theater, totally mesmerized by Elvis, the story and the scenery - all in living color. Elvis was singing and prancing around with a bunch of beautiful girls in bathing suits on the most exquisite beach I had ever seen. They spent the entire movie wrapped in sunshine, warm sand, blue water and waving palm trees.

I was hooked. I wrapped my scarf a little more tightly around my neck and tucked it into my double knit sweater as we burst out of the theater into a blustering howling wind. I pulled the hood of my duffle coat up over my head, leaned into that wind and silently and faithfully made myself a promise.

I am going to Hawaii.

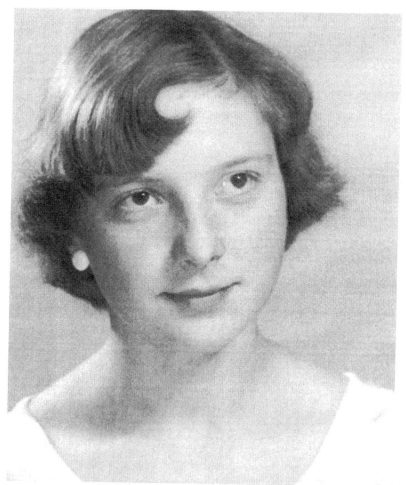

Gisela on her 16th Birthday

CHAPTER 10 -

NEW HORIZONS

My final exams at JAGS came and went successfully and I managed after much hassling and back and forth debating, to convince my parents that I did not want to go to University. This was not a very popular idea. In fact it was received like a death in the family! My stepfather had registered me at Oxford University on the day I passed my entry exam into JAGS. In 1952!

For the time being they blew me off feeling sure that by the time the summer was over I would have changed my mind.

I had not.

I hated to deal my dear sweet parents such a blow, but for the last three summers and the Christmas break at school – I had enjoyed a job. I had been a sales person at the 'Scotch House' in Knightsbridge - in the heart of London and one block down from Harrods. Every day I took the train up to Victoria Station and the bus to Kensington High Street. I really enjoyed the interaction with the public – especially visitors to the city from abroad. I loved all the beautiful wares and tartans from Scotland, and of course the paycheck at the end of every week.

On my lunch hour I would go out and explore the city around me. I would get on a bus and go a few blocks down the road, past Hyde Park, the Palace and Green Park to Piccadilly Circus. This is where I found all the elegant, beautifully decorated airline offices from different countries of the world that drew me like a magnet. I would spend much time looking in the windows of Air France, Alitalia and BOAC. Then, there, behind an enormous sheet of glass stretching from the pavement to the second floor was Lufthansa German Airlines. The summer sun would sparkle on the shining glass and sitting inside were three elegant ladies dressed in Lufthansa uniform and selling tickets to every corner of the world.

I was 'Alice' – staring mesmerized into the abyss of possibility down that rabbit hole.

Then, one day, at the beginning of the last summer before graduation, I had, with heart in my throat, opened that glass door and hesitantly, shaking with excitement and fear walked into my destiny.

It was a bold and reckless gesture but it was done, and I stood trembling in the middle of that beautiful office. A lady next to the window got up and welcomed me. Rescued me actually - and waved a cheerful hand to the vacant chair in front of her desk. A nameplate on her desk announced that her name was Mrs. Elizabeth Edwards. With shaking knees I sank gratefully down deep into the offered seat.

"I was wondering when you would get up the courage to come in and see us," Mrs. Edwards said gently while holding out her hand in greeting. "I have seen you many times walking by, because I sit right here close to the window, and you always spend a long time looking in. As you can see, we are all happy to have you visit us - and none of us bite - I promise!"

I felt the sides of my dry mouth lift gratefully into a smile. She was so very kind and warm, and spoke English with an accent like my Mother.

"I am so sorry to trouble you - disturb you," I stuttered, desperately trying to regain my composure. "I love looking in the window – your office is so beautiful and I enjoy seeing all the different pictures and displays you have from around the world."

"We are a 'round the world airline'. Are you interested in travel, my dear?"

"Oh yes, yes I am – very much," I heard myself say.

Ten minutes later that sweet lady had extracted not only my name but just about my entire family history!

Now she was leading me gently to the back of the room, under the stairs to the upper floor, and around the corner to a little office and kitchen. She went over to the stove, picked up the kettle spewing steam on the small burner, and poured the boiling water into the waiting tea pot.

"Come and sit here with me Gisela. We are not busy today, and I want to listen and find out a little bit more about what is on your mind, over a nice cup of tea."

I sat down across from her and gave her a quizzical look. I felt as though she was seeing right into my very soul.

"You are coming up to your school graduation soon," she said to me while passing a steaming cup of tea. "What are your plans – no – what are your hopes and dreams? Will you be going to university?"

"No!" I blurted it out so quickly that I accidently dropped a third spoon of sugar into my tea.

"I mean – my parents want very much for me to go to Oxford, but I do not want that – I want – to fly!"

My voice petered off into silence and my eyes hit the floor. It was the first time that I had put that thought into words out loud, and it felt funny to hear it.

To my amazement, my companion jumped up with excitement and clapped her hands.

"Yes Gisela – I can see you now walking in here some day in a Lufthansa uniform with wings on your chest. Perfect. Come with me there is someone I would like you to meet."

We went back around the corner and up the stairs to the first floor office. This was the Head Offices for Lufthansa German Airlines in the UK. They occupied a large office building around the corner on Regent Street but this was the 'jewel in the crown' ticket office on Piccadilly.

"Good afternoon Mrs. Warren," Elizabeth Edwards called cheerfully as we reached the top stair. "I have a young lady here that I would like you to meet and maybe we could introduce her to Herr von Frankenburg if he is not too busy."

A small lady in a dark business suit and bright white shirt disentangled herself from the papers on her desk and came towards us with outstretched hand.

"Good afternoon Mrs. Edwards. This is an unexpected pleasure."

The atmosphere and words were formal, professional and business like – very German – but the mood was joyful and welcoming.

"This young lady is Gisela Wheeler and she would like to know how she can become a Lufthansa stewardess," Elizabeth Edwards announced happily while shaking Mrs. Warren's hand warmly.

"Oh, come right on into my office Fräulein Wheeler, we can have a chat and I can tell you all about it," a tall, well-built very jovial man said from the open office door to our left and handed Mrs. Warren a stack of papers.

"Herr von Frankenberg you have a two o'clock appointment which is in twenty minutes. I can talk to Fräulein Wheeler and leave you free for that without any problem," his personal secretary announced with a worried look at her watch while adding the papers to the pile on her desk.

"Ja, Ja Frau Warren – keine Angst! You can let me know when Herr Doktor Hartigen arrives for his appointment, but in the meantime Fräulein Wheeler and I can have a nice chat," and with that he held out his hand in greeting and had me follow him into his spacious office decorated with a large German flag and model airplanes old and new.

That unexpected, lighthearted conversation took place completely in German, and would prove to be the door that would open many more thereafter in my life.

Herr von Frankenburg – the Director for Lufthansa German Airlines in the UK – told me he liked me and thought I had the world at my feet – I just had to learn how to 'plug' in to that fact - and believe it. He liked the fact that I rode horses and played hockey – he did both also, and missed his horse which had to stay in Hamburg while he was on assignment to London. He told me my school background was the best, and would help carry me into a successful future. However, at my graduation I would still be too young and had much to learn before I could join that special and talented group in the air, as a stewardess. To apply to be a member of that elite team I would have to be twenty years old and be fluent in at least two languages other than German. The languages were no problem but my four year wait was. Seeing my dejected, deflated look, he took pity on me and came up with a great idea. He asked me if I would maybe like to join Lufthansa as a clerk in one of the London offices. Meanwhile I could gain many skills while transferring through all the different departments and learn all about the airline. This knowledge and experience would be of great benefit later on when I was able to fly.

My face lit up like a Christmas tree. Of course I would love to do that! As a result we made a pact and shook hands on it as I left his office, promising to come back and talk to him upon my graduation from JAGS.

The afternoon dragged by and I found it hard to concentrate until my five o'clock release. I stood impatiently waiting for my bus to Victoria - I was so elated and flying on cloud nine all the way home. I could not wait to fill my family in on what had transpired that day and my good fortune.

I was going to fly - and Lufthansa was going to pay me to go to Hawaii!

I just had to wait four years.

New Horizons

CHAPTER 11 -

A DISTRACTION

My life and my weekends were very busy now. Especially now. School was done. I had started working for Lufthansa at the Piccadilly office as an assistant to Mrs. Warren – and I had a boyfriend.

His name was David Cianfarani. I had known him and played tennis with him for some years. He had gone to Dulwich College and belonged to the same tennis club I did. Basically, we had grown up together. His Mum and Dad were Italian and owned an Italian ice cream parlor in Herne Hill. In spite of my new life and all my new future plans, he demanded much of my time. He also made it his business to involve me in plans with other people and family, which I found hard to break. He had become so much a part of my life for so long now, and we were very close. He too was an only child – unusual in an Italian family, and he was like a brother. At least so I thought.

My Mother stopped by at my Lufthansa office one Friday afternoon on her way home from shopping at Harrods. I introduced her to all my new friends and showed her round the office. I introduced her to my boss and mentor Herr von Frankenberg. They hit it off immediately, happily reminiscing about Hamburg and the good old days before the war, while shamelessly indulging in a glass of German wine and beer. On the way home that night on the train she told me how proud she was of me. How much she had enjoyed meeting everyone and what a comfort it was to her to know that I was safe and in good hands for the future. I was blown away. She had never told me such a thing before. Ever.

That night Dick and Betty came over for dinner and my Mother had a great time telling them all about her afternoon at Lufthansa.

There was only one problem, she decided, and turning to me said, "It all made me feel very old – and I don't think you should call me Mummy anymore!"

I was incredulous. I knew she was making a joke but it felt strange. I looked at her in total bewilderment.

"But you ARE my Mother - what in the world shall I call you – 'Fred'?!?"

Of course, everyone laughed and raised their glasses

in a toast -"To Fred." Unbelievably it stuck and she was affectionately called "Fred" by everyone – except me - till the day she died. I always referred to her as that when talking to people about her, but I actually never called her Fred myself, until after she had died.

Six months later I was moved to the Regent Street office where Lufthansa had their UK Marketing, Cargo, Personnel and Accounting offices - everything behind the scenes for a working airline. I was to rotate through each department over the next year, to give me a really good overview and understanding of my new company. I enjoyed it and learned a lot.

One Monday morning there was a message for me on my desk when I arrived for work. Herr von Frankenberg would like me to stop by his office at 2 pm sharp, that day. I was a nervous wreck all morning – not able to concentrate on anything and being really distracted as the hours dragged by.

What could he want?

It was one thirty when I walked into the beautiful, busy fish bowl of an office on Piccadilly – waved at Mrs. Edwards and climbed the stairs to Mrs. Warren's abode. She greeted me with a warm smile and a handshake.

"Guten Tag Gisela – wie geht es Ihnen? Have a seat in your old spot across from me and tell me how is everything going? Herr von Frankenberg will be right back he is still at lunch."

I sat. It felt good to be back. Mrs. Warren told me

that she really missed me. I had been company and a great help. We continued to talk until suddenly there was a loud commotion downstairs and much laughter and merriment. The free standing metal stairs started to shake and the large, bulky form of Her von Frankenberg appeared accompanied by a gentleman in Lufthansa uniform.

"Gisela, Grüss dich – come into my office I want you to meet my friend Herr King from the airport. He runs our Lufthansa Heathrow Airport organization and he tells me we are getting very busy with more flights planned. He needs some help. Do you think you might like to work with him at the airport taking care of passengers? He has two ladies working with him in 'Traffic' and two people at the ticket desk, but he needs more help in 'Traffic' with the airplanes and passengers."

I found myself unable to speak – my brain blown away by his offer! I would need to go to Hamburg for a three week training period, and while there be fitted for a Lufthansa uniform. My heart exploded with happiness. Yes of course I want to go. I swallowed – my mouth dry. Next step - MY WINGS I thought!

The three weeks in Hamburg were wonderful and very enlightening. I learned everything I needed to know about Lufthansa and the workings of a large airport from behind the scenes. It was also the main crew base for Lufthansa German Airlines, and because I was underfoot and in the same 'space' as many flight crew members while I was in training, I was able to interact and therefore learn, a lot from these crew members. They took me under their wing when they heard that I was planning to apply as a stewardess with them in another two years, and were very patient and helpful with the multitude of questions I plied them with, before the end of my training and my return home to London.

Concentrate Gisela!

I dragged my mind back with difficulty to the here and now and the challenge of the questions to be answered in my end-of-course exam. I had to do really well on it if I was going to wear the uniform and represent Lufthansa at London's Heathrow airport.

Once back in London I had to figure out how in the world I was going to get to Heathrow every day for my early morning shift starting at six o'clock in the morning. We had recently moved away from Burbage Road to another London suburb further out called Bromley in the county of Kent. It was almost thirty miles from the airport as the crow flies. Getting to Piccadilly from Bromley was no problem – it was a one hour train and bus ride away – but Heathrow was another thing entirely.

One evening after returning home my Mother and stepfather had a surprise for me. They were watching for me as I walked up the garden path. When I came in the front door they turned me right around and led me back out to the garage. There, sitting cozily next to our new Jaguar was the shiniest, most wonderful looking Lambretta scooter that I had ever seen. I looked at my dad in questioning disbelief.

"For me?"

"Yes."

For the next two years I got up at three thirty every morning, left the house at four and drove a scooter all the way across London to the airport in Middlesex. Two hours. Rain, shine or snow. I rode in light blue leather pants and jacket against the cold, boots and a white crash helmet. I changed into my uniform and heels at the airport. After our last flight of the day was airborne for thirty minutes, to ensure it would not return – I drove the two hours back home. Days of twelve or more hours could be daunting – but not for me. Life was great - I could not have been happier.

David of course, was not. Our time together shrank because I now worked weekends also. We would see each other, sometimes only fleetingly, on my two days off. So it happened, on a day after we had played tennis, he told me his Mother had invited me to dinner after the game. It was a Wednesday night – my days off were Tuesdays and Wednesdays that month, and I was not too happy that this was sprung on me without prior warning. Now my days started very early in the morning and I needed to get to bed in good time in order to be viable and alert to drive 26 miles across London on a Lambretta scooter before dawn and in any weather. Unfortunately, I found it to be impossible to say "No" to his Mother. She was very sweet and kind to me, but she was very demanding and somewhat intimidating. She was very much the Matriarch of the family and ruled it with an iron fist.

"David, that is very kind of her, but you know I have to have an early night – I work tomorrow."

"Yes, I know you do and that is fine. She has planned an early dinner and you will be home in good time, I promise."

We walked up Herne Hill to his house holding hands while pushing his bike and carrying our rackets. We were very comfortable together – like two old shoes. I was seventeen and he was nineteen.

We started dinner at David's house with "Grace" which was alien to me – I did not grow up in a religious home. Dinner was delicious as always. Italians love to eat, and I love Italian cooking. The atmosphere was warm and the mood was very jovial. I ate too much and enjoyed a glass of wonderful Italian wine with dinner, resulting in me having a bit of a 'buzz' on afterwards. After dinner David took my hand and led me into the very elegant living room to play some music. He seemed a little strained, distracted - had my stress rubbed off on him?

"I bought a new record I want to show you. It is a surprise – I thought you would like it."

He handed me a flat paper bag with a new record in it. I looked at him sheepishly and pulled it out.

It was Frank Sinatra singing "Come Fly With Me."

I squealed with glee and flung myself at him. He grabbed me, lifted me off my feet and whirled me round and round as we laughed and hugged in delight. Then he put me down, took the record out of its colorful cover and placed it on the Victrola. The music started and 'Old

Blue Eyes' started to croon his way through the songs. David took my hand and pulled me to him and we started to dance. It felt so good and so very comfortable. I felt my whole body relax as I melted into him. We had never done this before. Been this physically close. He had never even kissed me. It felt so natural, and good.

Then, halfway through the record, he turned the music down and said, "Come sit – I have another surprise."

I walked to the sofa and sat as he had directed, wondering what else in the world he had up his sleeve.

He went to the beautiful, ornate antique French desk in the corner, opened a secret drawer and pulled out a small bag. He came slowly back to the sofa and sat down next to me untangling the ties on the bag. He was really serious now, his dark brown eyes showing no signs of the merriment we had just enjoyed.

"Gi – I know this is sudden. I know it must feel like coming out of left field to you, but I love you and cannot imagine life without you. I – my family – want us to be together forever. Will you please marry me?"

I sat, frozen, in total disbelief – panic rising slowly from the pit of my stomach – my eyes searching his face for an explanation. He could not be serious. We have had a long relationship. Yes! Close. Yes! But 'close' like family – like a brother – not like a husband or lover. I jumped up from his side like a scalded cat. He joined me on my feet and took my hand again.

"Gisela please hear me out. My Mother had this ring specially made for you from stones that have been in the family for 300 years. I know this is sudden and you don't have to answer now - think about it, but please - please - say yes."

Lufthansa – 1960-1962

My First Lambretta – Putzi I

CHAPTER 12 -

GROWING PAINS

Suddenly, my wonderful stable world, full of fun and promise, started to disintegrate. In the days that followed David's proposal and while still reeling and trying to work my way through such a preposterous idea my parents dropped another bombshell for me to deal with. They informed me they were separating and my stepfather would be moving out.

Arthur had been acting weird for some time now. I blew it off as mid-life crisis. He and I had been having some personal issues off and on over the last year. He would invade my space constantly and unannounced when I was in the bath or dressing in my bedroom. He would have his camera and try to take pictures of me by bursting in on me during these private times. Also, his 'fatherly' embraces had become way too intimate. I was very uncomfortable but did not want to worry my Mother, so just locked my bath and bedroom doors at night, and kept silent. I was reluctant to talk to my Mother, not knowing quite how to approach it. Did she know? It was hard not to know. I had been aware that she had seemed sad and withdrawn for some time, but when asked what was wrong she always made light of it and insisted nothing was wrong. She was just tired.

As a result of the tension between Arthur and I, I had made the decision a while back that the best thing to do was move out. I started looking for a bed sitting room for rent, close to the airport, but I was worried about my Mother. I did not want to leave her while she was so 'down' and pain and tension hung over our once happy home.

Then came the separation announcement.

About nine months earlier my Mother and Arthur had gone to a very fancy ball in London at the Royal Albert Hall. It was a fund raiser for a local research hospital and a lot of very prominent, wealthy people were there. They made a very dashing couple when they left the house - he in black tie and my Mother in evening dress and diamonds and fur. A perfect couple out for an evening of fun and splendor.

It seems they met a doctor and his wife at the ball that they really liked, and started to socialize with them many times since. Sadly, Arthur and Mary Wallace became an item over time and my Mother could no longer stay silent. Years later I would think back and decide that he had cheated before - 14 years ago - and history is just being repeated.

Now, David was pressing me for an answer. I could not give it. I was very fond of David and loved being around him. We enjoyed a lot of the same things and had the same friends – but marriage – did I love him? How would I, could I, possibly know? We had never had any kind of close or intimate relationship and I had related to him only like a brother. Also, what would such a thing do to my plans to fly with Lufthansa?

His mother inadvertently helped me out. She was planning a large engagement party and had a long list of family and friends on both sides of our families in mind. While at dinner with the Cianferanis one night, she explained that she was in a bit of a quandary. I was uncomfortable with the conversation as it was, because I had not accepted David's offer, so I had been silent through dinner. Now I looked up enquiringly as she dropped a bombshell.

"I am having trouble with the table seating arrangements Gisela, because, because – well, you know, you have not been baptized."

"Ex - excuse me?!" I stuttered – feeling sure I had not heard correctly.

"Well," she continued smoothly, "As you know, all our family is Catholic, some coming over from Rome and Palermo for the occasion – so exciting. As a result some of our older family members will have trouble eating at a table with a - with a – well you know – a 'heathen'. Of course the thing to do would be for you to become a Catholic because you would not be able to get married on the High Alter anyway. The service would have to be at the back of the church which is very sad and bleak and David is so very special, you know, being an only child. You could come with us next Sunday and meet Father Joseph at the church and he could help you get settled in a class. Then when you have become a Catholic we can have a beautiful service and all our dinner seating problems would be solved."

The world started spinning – I felt my dinner coming into my throat and I jumped up from the table and ran for the bathroom.

David got permission from his father to drive me home in the Mercedes. He too was upset, and did his best to calm me down. He understood my confusion and pain because I had twice refused to accept the ring his Mother insisted I take home with me. Twice I had refused. She was so very proud of it and its size – felt sure my parents would be impressed. I however, was not. I had so far left it sitting in its beautiful box on the table, much to her displeasure. The ring was supposed to convince me to become a Catholic?! That was going to be the criteria for our marriage?

EHHR – NO!

I paid a visit to my Vicar at the school Chapel on my next day off and explained my predicament. He was very sympathetic and realized that I had really thought this through. I did not wish to be a Catholic.

He checked the calendar and told me I could come to a nine week class starting next Monday. I thanked him and told him I would be there. He smiled and took my hand.

"It will not be easy Gisela. It is a class of 'Baptism for Those of Riper Years.' There are four souls registered for that class and you will be the youngest – they are all over eighty years of age!"

I smiled back – thanked him, and ran home. An enormous load lifted off my shoulders.

I confessed to David three days later that I could not do this. He was very important to me and I wanted to be friends with him forever – but I could not marry him. I loved him like a brother, but that was not the kind of love I should have for a husband. Love - I had no idea what that should feel like, but I knew that whatever it was – I did not feel it for David.

So this milestone in my life was over. Now I had to deal with two more.

Arthur moved out and my Mother divorced him. He then married the newly divorced Mary Wallace. It was a ginormous heartache for both my Mother and me. Who could have believed that we would move from a wonderful, happy family into shattered lives for both of us, within a year's time? We felt abandoned – suspended in time. Lost.

Dick and Betty were lifesavers. They were horrified

at Arthur's behavior and were rocks of reason, support and stability to lean on – especially for my Mother. We moved her into a smaller flat in Streatham – closer to them at Trinity Rise. I found and moved into a rented room in a house owned by a widow in Hounslow, Middlesex next to the airport.

So this is what it felt like to grow up.

Yes - and you will survive and become stronger for this heartache. You are now an adult.

Act like one.

CHAPTER 13 -

AN ALTERNATE OPTION

During my two years at London's Heathrow Airport in uniform with Lufthansa German Airlines, I worked very closely with British European Airways (BEA) personnel who handled all of our baggage, freight, and other ground based issues. It was standard practice for the home-based airline to do this for all foreign carriers.

One evening when I got home after a long fourteen hour day full of airplane delays and passenger problems, my roommate Sheila put on a kettle for tea and said casually, "Gi, you are wearing yourself out – today was tough for everybody but Lufthansa got the worst of it because of the rain, the fog, and then the mechanical delay you had. BEA is desperately looking for stewardesses. BOAC too – why don't you apply." It was a statement, not a question.

I looked at her sleepily over the top of my steaming mug of life restoring tea. I had kicked off my high heels and thrown myself gratefully into the easy chair next to the comfort of her cheery hearth. I said nothing. She looked at me smiling and added more coal to the fire.

With the flame restored she said, "I know, I know – that is not the plan, but BEA is desperate with all the new airplanes coming and you still have another year to wait to apply for flight duty with Lufthansa. BOAC is the same as Lufthansa, you have to be twenty-one. With 'us' you only have to be twenty – it can't hurt and would give you a head start later."

I did not respond but let the idea roam around in my weary brain.

On my next days off I drove to Streatham to visit my Mother. She was doing well and adjusting to being alone. Since Arthur's departure she didn't have a car. It really was no problem because she had buses and trains all around her going in every direction. But life was very different now and took some getting used to.

"Fred" had become very good friends with an old friend of the family, also a fine art dealer, who had a large gallery just around the corner from W. Wheeler & Son in St. James's. Clifford Duits was a widower having lost his wife to cancer the year before. He was horrified when Arthur left my Mother, and was a great help settling her in to her new home. He had three grown children and lived in a very old rambling house in London about two blocks from Harrods, on Wilton Place in Knightsbridge. Although he lived in town, he would make a point once a week of driving out to Streatham to check on my Mum and take her out to dinner, which she really enjoyed. He was charming, had a wonderful sense of humor and was great company. His family was Dutch. He loved history and loved to read. Reading hieroglyphics was his hobby of choice! He was very well-traveled and interesting to be around.

Clifford had called and left a message for me with Mrs. Weatherby. It was a necessary criterion to have a telephone if you wanted to work for an airline. You always had to be able to be contacted – even when not on duty. It took me a while to find such a residence because not everyone had a phone in those days.

I read with great glee the note Mrs. Weatherby had taped to my room door when I returned from work later that evening. Clifford said he would be coming to see my Mother on Wednesday and would love it if I could join them for dinner. He was hoping that my days off coincided with his visit and I would be able to come. They did - I had Wednesday and Thursday off that week.

It was a bright sunny morning when I packed my little overnight bag and slipped it on to the back of the Lambretta. I dressed myself in my aqua leather jacket and pants, strapped my white crash helmet on my head, adjusted my boots and set off on the long chilly ride across most of London, to Streatham.

An Alternate Option

I had a fun day listening to all my Mother's news and catching up. It had been a while since I had seen her although we kept in touch by phone. That was another one pound a week I paid Mrs. Weatherby - for the use of her phone!

Betty and Dick called and invited us over for lunch the next day – it was Dick's birthday and he had the day off. If the weather was nice we could take a walk around Brockwell Park in the afternoon and then get a bite to eat at our favorite Indian restaurant down in Herne Hill. Life was good; my world was starting to fall back into a comfort zone that had eluded me for some time.

Dinner out with Clifford Duits was fun, stimulating and delicious. Halfway through the entree and my second glass of wine, he dropped a bombshell.

"Gisela I think you are probably outgrowing that Lambretta by now. How would you like to upgrade to a small car? I have a friend who has a Messerschmidt for sale and I think it would suit you to a tee." I was stunned, my eyes meeting his over the top of my wine glass.

"Are you serious? Oh Clifford that would be amazing. Yes, this is my second Lambretta, you know. I had an accident early one morning last year, on my way to the airport, and wrote the first one off. That was a 150cc and I graduated then to this one which is a 175cc and is still almost new and in really good shape."

The larger Lambretta I had graduated to was great but very heavy, and truth be told it was too heavy for me to handle and I had some difficulty with it when I had to stop in traffic and at lights. It also was no fun to drive in inclement weather which is a way of life in England and was a factor in my accident. The car that hit me did not see me in the driving rain at a dark street corner and at four o'clock in the morning.

A Messerschmidt – what fun. Gasoline is expensive in England and also was in short supply in the late 1950's. Lloyd Messerschmidt was a famous man during World War II in Germany. He was famous for designing and building amazing airplanes for Hitler that almost helped swing the war in the wrong direction. Now the war was over and he was not allowed to build airplanes. So, being very inventive, he got around that problem by designing and building little cars that LOOKED like airplanes and were very cost efficient to run. They operated on a two-stroke mixture of oil and gas just like my Lambretta. A friend of mine had one and it was a kick to drive. It had three wheels and looked just like a small airplane with a heavy perspex glass cockpit lid, which lifted up to get in and out. Two people could fit in it driving one behind the other. The engine ran with three gears going forward and then to reverse you turned a lever the other way - behind the passenger seat - and WOW, you had three gears going backwards!! A Messerschmidt!! YYEESSS – what fun – I was hooked!

I applied to both BEA and BOAC (British Overseas Airways Corporation), for the open stewardess positions, and two weeks later I went for an interview with BEA. BOAC did not respond until I had interviewed and accepted the job with BEA. Sheila had been right in thinking they were desperate. One week later I was in class and became a "Seven Day Wonder." Class One was called this because they crammed four weeks of training into one week and - VOILÁ - we were graduated and flying the line!

My Messerschmidt Causing a Stir

CHAPTER 14 -

UNFURLING MY WINGS

My departure from my job at Lufthansa was sad and very difficult. I had such a wonderful time with all my Lufthansa family. They had taught me so much. In their tender care I had grown in many ways due to long hours, hard work in all weathers with a multitude of different passengers, crews and situations. Thanks to them I had gained an exceptional background in how an airline is run having experienced the good the bad and the very ugly in passenger handling. We had been through a lot and we were – family. I promised to stay in close touch and let them know when I was transferring to Lufthansa's In Flight Service.

Life as a flight crew member with British European Airways, or BEA as they were called, was exciting, exhausting, exhilarating, and challenging - but Oh, so much fun. I drove my little Messerschmidt to the airport and left it in the crew parking lot where it drew much attention. Now I was able to dress in my BEA uniform from home, instead of always having to change into my uniform upon arrival at work on my Lambretta. My airport background with Lufthansa was a great help to me upon my transfer to BEA as a stewardess – especially because I was 'A Seven Day Wonder!' It gave me a terrific heads up over my other classmates that had not had any airline experience before, as well as being a great tool for teasing by my more vintage crew comrades. They loved this and made much use of it, at my expense.

BEA had received six brand new Vickers Viscount aircraft from the factory shortly before my arrival – hence their desperate need to hire more flight crews and the very short, hasty stewardess training. The aircraft were very sleek and modern with four big Rolls-Royce engines able to carry sixty-six passengers and four crew members. Our flight routes were varied, encompassing most large cities of Europe and reaching as far as North Africa. From there British Overseas Airways Corporation – otherwise referred to as BOAC – would take over and continue with flights around the world.

I now had a whole month of blissful flying under my belt and was feeling very comfortable with the airplane and my on board duties. There was a galley in the back of the aircraft and we served sandwiches or snacks of some kind or other, on every leg. If the flight was more than two hours in duration the meal was sometimes hot. In the early 1960's on that airplane we had no carts yet, and hand carried everything. The flights were seldom full, usually ranging anywhere from twenty to forty passengers depending on destination. London to Paris we usually had more, and if we were flying to Spain or Portugal we had full loads – people from rainy old England escaping with great delight to warm sun.

Today I was flying to Gibraltar with a steward called Charlie Pekham. I was very excited, being a bit of a history buff. I had learned about "The Rock of Gibraltar" at school and found its history fascinating. It had proved of priceless value to the British and the Allies during the war. Now I was actually going there with a long lay-over before heading to the island of Malta and on to Tripoli and Benghazi. I had brought a history book on Gibraltar with me in my bag, and planned to read up on it on my lay-over.

It seems that the six square kilometers that is The Rock of Gibraltar, has been a prized site for centuries, and its people have witnessed many battles and sieges over time. Its position guarding the entrance to the Mediterranean is unrivaled and has been fought over by Spain, France, and Britain all of which claimed possession at one time or another. Britain finally succeeded and after much fighting and trauma solidified its victory in 1703 with the Treaty of Versailles in 1783. Spain almost managed to retake The Rock in 1779 creating The Great Siege of Gibraltar. As a result many tunnels were dug in The Rock for protection thus creating a new city underground that connected with natural caves in the rock itself. These tunnels are a feature and a legacy that came in mighty handy during World War II. Eisenhower made this his command post when he needed to protect the narrow Straights of Gibraltar leading to the Mediterranean and assist the fight against Rommel - The Desert Fox - in North Africa.

The Rock had also played a part in British history, when in 1803 Lord Nelson's fleet had visited The Rock after victory in the great Battle of Trafalgar. Many men had died and the Trafalgar Cemetery was created there to bury some of the dead. Lord Nelson himself had lost his life during the battle and his body was embalmed and brought ashore in a cask of wine to await transportation to England and burial.

I was blown away by all this history and could not wait to get there and check some of it out. I especially wanted to climb The Rock and get up close and personal to some of the famous Apes that are wild and in residence there. The Apes are steeped in history and superstition states that if the Apes leave, the British will go as well. Apparently the Apes got so very excited and VERY noisy if a ship approached the coastline that it tipped the residents off that danger was imminent. Thus, the story is told in Britain, that Gibraltar was so valuable and necessary to the Allies, that Winston Churchill made sure the Apes were protected. So much so that their numbers were replenished with more Apes from Africa. Just to be sure! He did not dare risk getting 'jinxed.'

So much history, and now I was getting to check it all out for myself. I could hardly wait to start this adventure.

Our Captain was Geoffrey Padget and our Copilot introduced himself as Tristan Walker. Captain Padget informed us that we would encounter some turbulence over Germany but it should not interfere with our meal service. The flight time was three hours and forty five minutes. He liked his tea hot and black and please "keep it coming." After the short briefing we all four walked across a windswept tarmac to the airplane sitting silent and ready for our arrival.

Charlie and I stowed our bags, checked our emergency equipment and counted our meals in the galley. Both back doors of the airplane were open and the cross ventilation felt good. We had fifty-eight hot meals and five crew meals. The captain and copilot were not allowed to eat the same thing, so we always had spare meals. The two liquor chests were still 'custom sealed' so all was well and we were ready. Charlie went up to the cockpit to inform the Captain of that and to give the Captain his hot black tea which we had brewed upon arrival. Tristan wanted a can of Ginger Ale which was easy enough, and as Charlie came out of the cockpit the passengers were being led across the tarmac.

The "Traffic" person from BEA leading the passengers to the airplane was no other than my roommate Sheila. She came bouncing on board and we hugged and greeted each other with great glee.

"You have forty-nine passengers, a wheelchair passenger and an unaccompanied minor – this is all her paperwork – her mother will meet the airplane and take her off your hands in Gibraltar."

Charlie took over the wheelchair passenger getting him securely seated in the front row where he had lots of leg room. I took the little unaccompanied girl's paper work from Sheila, reached for the child's hand and lead her to a seat in the back near the galley. We settled the passengers and their bags and closed the aircraft doors. Charlie did a head count and reported to the Captain that we were secure with forty-nine passengers and an unaccompanied minor. He asked the pilots to please have a wheelchair meet the airplane in Gibraltar. Charlie slammed the cockpit door and started his Welcome On Board announcement while I did a cabin check for seat belts fastened and baggage stowed. By now the engines were all running smoothly and we left the gate right on time. Then after one more quick cabin check on my way to the back of the aircraft, we took our seats – Charlie by the forward door and me aft. We were on our way to "The Rock."

After take-off we did a beverage service. This always took a while because we had to hand carry everything on small trays. Next, we started the meal service. As the Captain had predicted we now had some slight turbulence which made walking up and down the aisle with three meal trays a tad tricky to say the least. Dinner today looked really nice. We were serving chicken with rice, a small salad and custard pudding. I had served about three quarters of the cabin and the passengers all seemed happy and ate with gusto. I had just turned out of the galley with a hand full of trays when the aircraft lurched, my feet left the floor and I became airborne. The result was ugly. I ended up in a gentleman's lap with three trays of food raining down around us. He and I both had chicken and rice in our hair and some of the salad ended up in his gin and tonic. He and I were a mess to clean up but he was a frequent traveler and was a good sport about it all. The rest of the flight was calm and uneventful.

Upon arrival and after the passengers had deplaned the captain came out of the cockpit took one look at me and said with a start.

"Gi whatever happened – did somebody throw their lunch at you?" I gave him a wry look and said "No, YOU did, when you ran us off the road into a ditch!" His face creased into a wide grin and he hugged me as we went down the aircraft steps to the crew bus. "I'm sorry – we did not see anything coming on radar, did we Tristen?" He threw casually over his shoulder at the copilot bringing up the rear. "Otherwise we would have warned you. I'll buy you a beer when we get to the hotel to make you feel better, how does that sound?" He was still smiling when he slammed the bus door after throwing our crew bags into the back seat and we all settled in for the short ride to the hotel.

The mood was very jovial and upbeat and much of the mirth and merriment and teasing was addressed to me. Especially when Charlie told them proudly that I was "A Seven Day Wonder" and I had handled everything like a pro.

Then, when we drove out of the airport and joined the crazy traffic flow into the city, he turned to me and said, "Have you been to Gibraltar before Gi?"

Unfurling My Wings

I admitted I had not, but was very excited at being here and planned to explore the city tomorrow.

"That is wonderful Gi – I am meeting a whole group of friends tonight and I would love for you to come and meet them. You can join us and we can show you everything you need to know about The Rock of Gibraltar. You will love my friends, they are very colorful, fun and live here - we will have a blast." I thanked him profusely and he said, "Knock on my door at seven, I am two doors down from you, and we will show you the town. Wear comfortable shoes."

So that was how we left it. We checked into a really nice hotel right in the center of town. I dumped my bag on the large bed, tore off my still sticky, soggy uniform and jumped into the shower. At seven on the dot I let myself out of my door thoroughly rejuvenated and very excited to hit the town and explore The Rock. Charlie had a room two doors down and I was quite surprised to see a very elegant lady in high heels - beautifully dressed and coiffed - let herself out of his room and slam the door. I made a clumsy attempt to reenter my room so as not to disturb her, thinking I had caught her in some indiscretion with my steward. To my utter amazement Charlie's voice came out of this impeccable personage's body and said, "Gi you are so prompt – that's great let's go. Everyone is waiting for us." It was Charlie Pekham in total and complete "drag!"

I sped down the hallway after this amazing apparition. We tripped over our captain and first officer on their way to dinner in the spacious lobby. Both had big smiles on their faces and waved merrily at us as we sailed smoothly and with great momentum onward, and out into a warm, sunny evening.

The Eagle – though thoroughly startled - arose and spread her wings in gentle - if slightly hesitant - flight.

BEA - 1962-1964

CHAPTER 15 -

BERLIN 1962

After returning from Gibraltar, I came home to find my new schedule for the next month was quite varied. I worked several Glasgow, Edenborough, and Dublin trips, mixed in with some precious trips to sunny and warm Mediterranean countries. One of them was to the island of Malta, and I would be gone almost two weeks. It was an extended crew base for BEA. Instead of flying out of London we flew out of Malta every trip, all around the Mediterranean. It was wonderful and we came back to dark rainy London suntanned a golden brown and very healthy looking.

After my return, I had three days off and went to visit Fred. She was doing well and making new friends. I took her on a long drive out of London into the country in Putzi III – my little Messerschmidt. I had called my first Lambretta "Putzi" so this was Putzi III! It was a lot of fun and we caused much excitement and interest everywhere we stopped. We had a delightful lunch in the beautiful English garden of an old pub near Windsor castle, and I promised Freddie we would do this again next time I came to visit. It felt good to see her smile and happy again.

Upon my return home, there was a message for me from crew scheduling. They were sorry to tell me that they were reworking my schedule and taking me off my assigned line of flying. I was now scheduled to "dead head" - fly as a passenger – to Berlin. I would be gone for exactly six weeks so I had to ensure that I pack enough and for every eventuality.

British European Airways was in partnership with Pan American World Airways flying the Russian created "corridor routes" in and out of West Berlin. After World War II my old homeland had been divided into two parts - East and West Germany. Berlin was located in the middle of Russian controlled, Communist East Germany. Berlin too had been divided in half by a dreadful wall. BEA and Pan Am were the only airlines allowed to fly in and out of West Berlin along the Russian supervised corridor routes. As a result we had a crew and maintenance base there. The pilots were all British with a few Canadian pilots who had joined us due to heavy lay-offs in their homeland. The jet age had hit the

Americas hard and as a result not as many pilots were needed. The jets were larger but only needed two pilots as opposed to three in many propeller airplanes. BEA needed them to help us crew the six new Vickers Viscount turbo prop airplanes we had recently purchased in the early 1960s.

The stewardesses based in Berlin were all German however, and the airplanes and crews only flew to cities within Germany. Once a year everyone was required to return to the home base, London, for a week of recurrent training, or RT. This is basically an annual re-certification refresher course required for all flight crew. Hence my new assignment. Because I was fluent in German and had dual nationality I was very desirable. They had six German stewardesses in Berlin, so I would take over one week of flying from every one of these ladies in turn, so that they might go to London for their annual RT.

It was wonderful being back in Germany. I stayed in one of the best hotels in town, The Kempinski on the Kuhrfuesrstendam right in mid-town. I loved it and thoroughly enjoyed the change of pace and the short flights between so many German cities including Hamburg. The pilots were fun and very friendly and would invite me to come and sit in the cockpit for take-off or landing when the weather was good. We flew in and out of Tempelhof Airport which was an amazing feat, because it was right in the heart of Berlin and you could look into everyone's kitchen or living room. It felt like you were landing that big Viscount on a highway downtown!

Once, on a flight from Frankfurt to Berlin we were having a very bumpy ride and the captain insisted everyone stay seated including the stewardesses. I was sitting forward with my back to the flight deck and facing the passengers. Many of them were maybe flying for the first time and were looking very stressed and worried as the airplane kicked and swayed. A lady got up to go to the bathroom and promptly fell across her neighbor trying to get out of her row, as the airplane bucked up and down and sideways. Because people are not familiar with the environment that they are in on an airplane, they are inclined to do silly things. Getting up during turbulence is one of them. My fellow stewardess seated in the back of the airplane and I called out to her in unison, to please REMAIN seated. She untangled herself from her neighbor and sank back in her seat, face aglow with embarrassment.

Suddenly there was a change in the atmosphere and mood in the cabin, as all the passengers turned to look out of the windows left and right. I was sitting next to the door and could not see out so it was very strange for me to observe this from my jump seat – it was fear!

Just then the captain called me on the phone next to our station.

I picked it up and he said, "Hi Gi. Has anyone back there noticed that we are not alone? We have company. Because of the bad weather I was trying to find a smoother ride and veered slightly out of our 'corridor.' We are being 'Buzzed' by two Russian Migs."

"Oh so that is the problem. Yes, Captain, everyone is

looking at them. I was wondering what they were all looking at. Do you want me to make an announcement?"

"No that's OK – I'll do it, thanks." and he hung up from me, and promptly did so. The passengers absorbed every word, but his voice was very calm and he indicated – no worry - that this was routine. The Russians did not like us straying out of our corridor. He had apologized to them he said, they would be gone shortly. And they were.

The rest of the flight was calm and uneventful and the mood in the cabin became jovial with anticipation of our arrival. The captain turned on the Fasten Seatbelt/No Smoking sign and when we had descended below 10,000 feet, he called and invited me to come up to the flight deck for landing.

I was thrilled and buckled myself into the seat behind him. By now it was dark and I was amazed at the lack of light or illumination on the ground. Ahead of us was a bright blaze of light that was West Berlin - but between us and that bright halo – there was nothing. We were flying over East Germany and it was eerily dark and foreboding. Totally unlike the rest of the continent which was alive and bright and beautiful at night. Our descent took us past the city and around in order to land on the north/south runway. It was the strangest sight I had ever seen. The city was completely cut in half – communist East Berlin was bathed in inky blackness, completely dark, with just a few pinpoints of light scattered here and there, and West Berlin was a blaze of light and energy and life!

It was an experience never forgotten, and I remember

returning home to England with a whole different and slightly sad perspective of my old homeland. I thought back seventeen years and the deep sadness that I had felt all those years ago at losing my wonderful Tante Hennie and Onkel Jürgen and their beautiful farm, to a Communist regime - a country divided and cut off from civilization and family.

They, sadly, did not survive, but I did.

The Eagle is in flight and living on for all of them.

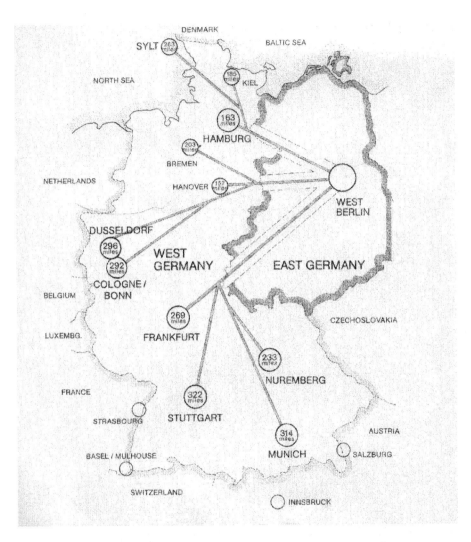

BEA Corridor Routes in and out of West Berlin

Berlin 1962

CHAPTER 16 -

MALTA 1963

The sun was hot. Burning in fact, with a slight prevailing wind fanning us gently as the crew of BEA flight 603 lay out by the pool at the Phoenicia Hotel on the island of Malta in the Mediterranean. We were all covered in baby oil mixed with iodine, and like typical northern fair skinned Europeans we were "sun worshipers!" We lay there like fish on a slab – slowly baking. Every now and then we would take a quick sip of beer to cool our baking bodies.

We had lived through a somewhat traumatic arrival and overnight experience and were happy to all stay safely close together for the time being. Our assignment was for a 10 day stay working out of our mini base on the island of Malta with plenty of 'down' time in between flights. We left London airport the day before in the late afternoon taking off into heavy rain clouds. The flight to Malta was uneventful until we were clear to land. Malta was having a bit of a problem with bats. BATS. They are a good thing in as much as they eat bugs, but they are inclined to hang about around the runway lights at the airport which attract bugs. That is not such a good thing. Our descent and approach was uneventful and normal until we were over the airfield perimeter and there - we hit bats!! They were everywhere around those runway lights – and we hit a gazillion of them – KERBOOM!! The aircraft shuddered and shook like a dog trying to rid itself of a bucket of unexpected water. Our captain, having experienced this before upon arrival in Malta, was ready, and so corrected the resulting yaw, and placed the airplane firmly on the runway. THUUUUUMMMPPPPP. That is the difference between a LANDING and an ARRIVAL. This was an arrival!

After clearing customs and passport control, we got into our waiting crew bus and drove around the perimeter of the airport to our hotel in Valletta. As we drove out of the airport gate we saw our airplane being towed to a hanger. It would need a thorough cleaning and maintenance check to remove any little gory bat remains from the engines and fuselage. The captain quipped, "I'll be sure to turn off our radar before landing next time I come, it must have interfered with theirs – so sorry guys!"

We checked into the hotel which was old, spacious,

and very glamorous. It had the old Casablanca movie charm. You could almost feel Humphrey Bogart leering over your shoulder in the lobby as you checked in to your room. I happily threw my bag on the bed – stepped out of my uniform and jumped gleefully into the shower. My room was very spacious with beautiful antique furniture and a wide open marble floor. After freshening up we all met in the lobby and went into Valletta for dinner. A pleasant uneventful evening with much good humored banter about our "arrival" in Malta followed up by an early night of cozily curling up on the bed with a good book.

I was in the deepest most blissful sleep of the night when I was rudely awakened by an unbelievable crash and a splintering of glass. I came wide awake and sat up in alarm, to find my bed – no my whole room shaking in a frenzy. It was the scariest sensation. I could see nothing. The room was dark, totally black. My room faced the pool and I had pulled the ginormous curtains shut before going to sleep. They hung from a twenty plus foot window to the floor with a double thick backing which would keep the heavy sun light out in the morning. When I sat up I could not remember where the window was. Where I was. It did not matter, there was no light and I could not find the switch to my bedside lamp.

I was totally traumatized as the shaking continued and my befuddled sleepy brain realized we were having an earthquake!! The crash and splintering glass was from the water bottle next to my bed hitting the marble floor. Then my clouded brain kicked in and screamed at me that the island of Malta was created by volcanic movement of the earth millions of years ago. The island is the top of a mushroom shooting up from under the ocean. What if the 'stalk' of that mushroom broke off?!

The shaking stopped as suddenly as it had begun. Logic started to creep into my frantic thoughts. I could not get out of my bed because the bed was against the wall on one side and I had a mass of splintered glass on the other. I crawled to the bottom of the bed and gingerly lowered myself to the floor. I was halfway across to the window to open the curtains when my phone rang. I froze. Should I go to the window and open the blinds to see if anything was happening, or should I answer that stupid phone that was sitting in the middle of my carafe of shattered glass.

The logic of my Germanic brain kicked in and the phone meant human contact. I slid my feet gingerly over to the ringing phone and picked it up.

"Hi Gi – just checking to make sure you are OK. Just a little earthquake – happens a lot down here - not to worry old girl, go back to sleep and I will see you at breakfast." It was my British captain dutifully checking on his crew while calmly delivering a normal English understatement of facts. I smiled in spite of my shaking legs, fished for, and found the lamp and turned on a blaze of comforting light. I stared at the shattered glass at my feet, decided I had nothing to clean it up with and so I gingerly walked around it and climbed back over the foot of the bed and down deep into a million soft pillows and slept.

Thirty-six hours later the airplane was given a clean

bill of health and we were to fly to Benghazi as quickly as possible. The earthquake had done terrible damage to some areas of North Africa. We were now rescheduled to pick up a band of refugees from out of the desert in Benghazi and deliver them to some emergency camp in Palermo on the Island of Sicily for temporary shelter.

We took off from Malta with an empty airplane on our way to an emergency mission of help in Benghazi. Our normal Benghazi, Tripoli, Palermo flight had been canceled in order for us to take care of this emergency. I got the feeling our captain seemed a bit distracted and quiet upon our departure, but did not give it further thought. We were headed for North Africa and worlds unknown to me.

The airport tarmac shimmered and seemed to float in the heat of the afternoon sun as we rolled into Benghazi on the northern coast of Africa. The steward working with me opened the forward door and a whooosh of hot air shot through the cabin. I, in turn opened the back door and a bunch of Arab ground crew ran a set of stairs up to the airplane.

"We can board right away, if you are ready?" the disheveled, scraggly agent dressed in pieces of a BEA uniform yelled at me. "We have sixty-eight passengers and five babies. No one can speak English except three French Nuns traveling with them." I gave him a thumbs up and then turned and ran up the aisle to the cockpit.

"Captain, they are boarding sixty-eight passengers and five babies – where do you want us to put the two extra 'strap hangers' – we only have sixty-six seats on this aircraft?"

A 'strap hanger' is the term we used for passengers on board an airplane that did not have a seat.

The steward that was working with me looked a little 'green.' He had mentioned he was not feeling too well. Some of the fish stew he had eaten for dinner had given him a bad night, and he had thrown up throughout the earthquake.

"I will go down to the back and take care of it Gi; you stay up here, do the announcements and get everyone settled." With that he ran off down to the back of the airplane as the first passengers boarded. I must say, I was in total awe as they came up towards me in the cabin. Something kicked in to my befuddled brain and I blocked off the three front seats in front of my jump seat. It was an almost totally unconscious gesture. I wanted the nuns up close to me in case I needed their help.

The passengers that came up the cabin were all men. Men in long wrap around robes covering them from head to foot. All were barefoot. Each had one piece of hand baggage, but hand baggage the likes of which I had never seen before. These people had been completely 'bombed' out of their desert homes by the earthquake, and they were carrying all that was left of their lives. These were people that had never even seen an airplane before, let alone set foot in one. They had no concept of what was happening to them as I herded them into one row at a time and sat them down. Thinking all was well it did not occur to me to check behind me as I moved down the cabin. Eventually I came across my first woman passenger. She came on board with a baby of about a year old. I seated her, but the baby somehow got away while I was trying to fasten her seatbelt, and disappeared under the row of seats in front of her. I turned around and caught my breath in disbelief. All the men that I had so far 'seated' were up and roaming all over the front of the airplane. They were poking and touching everything around them and then smelling their fingers. What that indicated to them I could not imagine, but then to my horror I saw one urinating up against my jump seat!

For a moment my world stood still. I bent down and calmly scooped up the baby from the aisle and practically threw it at its mother, telling her to hold onto it very tight. I then shoved my way through all these moving bodies to the front of the airplane and grabbed hold of the 'urinator' from behind by his voluminous robes and deposited him in a seat. Then I turned unhooked the PA and in my loudest voice announced to everyone in general.

"SIT DOWN!" I almost shouted this while making a motion down with my hands. The result was quite magical. Everyone sat – just not necessarily on a seat – they just sat. Now I started to pay attention to the baggage they were bringing on board. None of it would fit into the overhead bins. This aircraft was built before it was decided we needed a door to close the overhead bins. These bins were just glorified hat racks. I did the very best I could to store everything after the passengers were seated - in a seat and not in the aisle or in front of the entry door.

The First Officer came out of the flight deck and said

he was going to open the front door again because they were sending the three French nuns up that way, and Operations needed to bring the aircraft paperwork up when we were ready. I was so very happy to see these three wonderful nuns and escorted them to the front row seats I had blocked off for them. They were delightful – were familiar with these people, having lived among them for some years – and spoke quite good English. I knew they would be a great help on the flight to Palermo. Their large white head gear, however, could pose a possible problem in the tight space of the cabin.

After the nuns were seated I called Jason the steward in the back to see how he was faring. He was not happy. He had mostly women and also babies in the back by him. He also had another problem. Two, actually. The first applied to me and the second I could not help him with. The first was that he had three passengers that had no seats, the second was he had thrown up again and had terrible diarrhea.

I went into the cockpit for advice about the three strap hangers. I needed to know what the captain wanted to do about them. He decided we could not leave them behind, because they belonged to families on board. He said that I could sit in the cockpit for take-off and landing which meant one person could have my jump seat. One of the teenage girls in the back would have to sit on somebody's lap – that would open another seat, then with Jason sitting on his jump seat by the back door, the last strap hanger would have to sit in the back toilet. That would be our fix.

It took us almost two hours to get the group settled

and strapped in. They were loud, undisciplined and completely ignorant of their surroundings. We were already exhausted and we had not gone anywhere yet! The fastening of the seatbelts was a complete puzzlement to these people and I noticed that many that I had not had time to help – just tied them in a knot. Now the doors were closed and we were ready for take-off. The nuns were sitting in the front row smiling encouragement at me. The captain decided we should not make our emergency announcements because it would be futile. Our passengers would have no idea what I was talking about and it would just scare them. I did do an emergency briefing with the nuns. Explained about the life vests and were they were and how to put them on. Then I went into the cockpit and strapped myself in. As we were taxiing the crew call came in from Jason in the back of the airplane. The First Officer pushed the open speaker button and Jason's tired voice came over the air.

"Captain - err I just wanted to - eerr let you know that - eeerrr – uumm I will be in the toilet for take-off and landing. Not feeling too good - errr – I think it is the safest place for me right now. Hope to feel better anytime. The cabin is secure and the passengers are chanting, which is a good sign - I think."

The captain acknowledged the call, pushed the throttle open wide and we were on our way.

Malta 1963

CHAPTER 17 -

THE HANDS OF GOD

It took forever for us to get airborne. We had so very much excessive weight in the cabin and in the holds. We finally staggered into the air just a few feet before the runway ended in the ocean. The take-off was a tad white knuckle to say the least, and I sensed quite a bit of unease in the cockpit between the two pilots. I had no time to worry about it – I had sandwiches and drinks to distribute to my passengers, so I left and entered the cabin. The passengers were still chanting and swaying back and forth with the rhythm. I suspected that several of them had relieved themselves during take-off because some of the carpet was wet as I passed and there was a distinct odor of urine in the air.

Whether this might have been out of need or fear I will never know. The galley was aft, and I chose to continue on, uninterrupted to start the service. Jason was nowhere in sight. I started right away carrying as many sandwiches as I could get on a tray to the front rows, working towards the back. I had served several rows when on my way back to the galley I thought I smelled a strange odor. An odor I could not immediately identify, and I went on back to collect another load of sandwiches.

IT WAS FIRE!!

One of the passengers had unpacked his baggage sack, and set up his little butane stove and started to cook dinner! They were in a world decidedly alien to their own. I turned with well-trained automatic movement to the rear jump seat, grabbed the chemical fire extinguisher from its rack and shoved my way through the milling passengers towards the culprit. Needless to say my dousing of the little stove caused chaos and panic amongst that group as I emptied the fire extinguisher of its contents on everything within an eight foot area. The sudden loud noise brought everyone to their feet.

The nuns came running to my side and were a great help with the clean-up and settling the passengers back down; explaining to them as they did so that stoves MUST stay in bags.

I went forward and explained to the captain what I had just done. He looked at me in total disbelief.

"NO! How on earth did they get that on board! I wonder what else could be on board this airplane that

Ops 'forgot' to mention on the paperwork." He turned to the first officer and said, "Jim do go back and take a look and check things out, there's a good chap. Good job Gi – carry on both of you."

We left the cockpit together and I left Jim pouring over the mess in the cabin while I returned to delivering the rest of my sandwiches.

Jason finally emerged out of the toilet and helped distribute drinks – mostly water or hot tea, which was a big help. The cabin settled down and many of the passengers slept. He showed some of the men the toilet and explained how to use it, but most did not seem to understand and just stood in the open door and urinated on the wall.

We were about an hour and a half into our three plus hour flight. I had finally managed to sit down and relax with a cup of tea and talk to the nuns, when the tone of the engines changed. I looked up as the aircraft lurched slightly and the number four engine started to backfire, backfired again, and again, and yet again, and then the propeller stopped spinning and I knew it had been 'feathered.' The nuns looked at me startled when the 'crew call' came to the intercom phone by my jump seat.

"Gi we have a little snafu," the captain stated in his usual mode of 'understatement.' "We have 'feathered' an engine but all seems well so far. It will just take us a bit longer to get to Palermo. Do you need help with the passengers?"

"No captain all is calm so far. I don't think they

know what is different and I will do my best to act as if nothing is wrong." I hung up the phone and explained the problem to the nuns who seemed to trust what I told them and set aside their fear.

I went back to explain our problem to Jason who was washing down walls by the back door. He looked thoroughly drained and did not take the news in stride. I picked up a garbage bag from the galley and started to go back through the cabin picking up trash when the airplane bucked and swayed and the number two engine started to backfire and smoke started to billow out of the housing. The propeller stopped and I headed for the cockpit. The cockpit felt warm as I entered and the captains' blue shirt was damp with sweat as he and the copilot moved levers, changed switches and did their utmost to cope with the problem. The captain was pressing the intercom button to call us when he saw me standing behind him.

"Gi we now have quite a different problem and I do not think we will make it to Palermo. I am advising Malta and we will be heading there. We are about forty five minutes out and I think you had better get our passengers into life vests and prepare for an emergency." My heart was racing, beating in my throat. This could not be happening – not on THIS flight – with THESE passengers – and JASON. Where was he? The last I saw him he was disappearing back into the toilet.

My brain was churning. The world was spinning and it was spinning on my shoulders.

The aircraft shuddered, trembled, and shook as if in

pain. It was slight turbulence from winds coming up from the ocean as we got closer and closer to it. It was no greater than my pain. I lowered myself hastily into my jump seat across from my friends the nuns.

"I am going to need your help, your strength, your knowledge, and your God." I explained everything to them and then I asked if they were OK with that and would they please help me? They were solid as rocks and just needed me to tell them what to do.

I split them up - two to the back of the airplane to help the women and children – and one in the middle to work towards the rear. We went over the life vest information again. I donned my life vest and showed them exactly how it worked. Then I made them put one on which was hard with their large white starched linen Cornette on their heads. This done I sent them off into the cabin and told them ON NO ACCOUNT MAY ANYONE INFLATE!

I turned to start getting my passengers suited up when a small strong hand caught mine. It was Sister Adelaide.

She pulled me into her group and said, "Come, we pray."

The Lord's Prayer had never felt so good, so soothing, so calming as it did that day at 20,000 feet over the Mediterranean Sea - in French.

I went to the back of the airplane to brief Jason. He was still indisposed in the toilet. I pulled the life vest from under his seat and hung it on the toilet door knob and told him where it was. Then I went back up to the front to start my work of finding a life vest under each seat of these poor simple men and tried transitioning them into the 20th century. Sister Adelaide was at my side but we had the hardest job convincing these men that they cannot pull the tabs until they were OUTSIDE the airplane. The cabin was a zoo. These sweet innocent passengers had kicked in to their gentle brains that we had a problem. The first clue was that two of our propellers were no longer spinning. The result was not pretty. My three nuns were without a doubt, my strong, calm lifesaving fix. It was thanks to them and their calm business like attitude that helped get the cabin back under control. The nuns were also well known and trusted friends to all of my passengers.

I finally entered the cockpit with no time to spare and to inform the captain that the cabin was ready and to await any further instructions. I was horrified to find how close we were to the ocean. I had been so busy in the cabin I had been unaware of our descent, Through the cockpit window I could see land – I knew it had to be Malta but it still seemed so far off.

"Right Ho Gi – go strap in and get ready – I will call you on the PA with instructions and when to brace." I turned to go out of the cockpit door and gasped at what I saw.

The entire cabin was one solid mass of moving yellow jelly - every vest had been inflated No one would be able to exit this airplane in the event of a ditching. I turned in horror back to the captain and stepped aside so he could view the cabin. He unstrapped his shoulder harness so that he could get to his feet, pulled a sharp pointed knife out of the side of his boot and thrust it into my hand.

"Take this Gi - work your way down the aisle and deflate one side of everyone's vest as you pass – best we can do. We are about six minutes out – I think we can make the beach. GO!"

I did what he told me but did not have time to go back up front before his voice came firmly over the PA "BRACE FOR IMPACT!" I slid myself down between the last row of seats and covered my head.

We were indeed in the hands of God.

The Hands of God

CHAPTER 18 -

TRIPOLI 1963

Three months later I was once again scheduled to fly the 'Ten day Malta' trip. I regarded the scheduling sheet in my hands with some trepidation. I loved Malta and the flying down to the warm lands of the Mediterranean – BUT, could it all happen again? We got lucky that time - we took out some of the approach lights at the tip of the runway - hit the runway – HARD - BUT we made it to safety with not an inch to spare!

Prayers were answered that day, indeed.

Now I am to go back.

I had made a mental note of the aircraft identifier, G-ALWF, should I ever encounter it again in my travels. Even though I knew intellectually that the aircraft was not to blame for our plight that day. IT WAS BATS!!

I packed my new bikini and thinking only positive thoughts drove Putzi, my little Messerschmidt, to the airport and parked. The briefing room was crowded with crews coming and going and I waved and went to talk to a German girl that I had flown with in Berlin. She had spent her days off in England and was heading home. Then I went to find my crew and was delighted to find that my friend Charlie Pekham was going with us. I had thoroughly enjoyed him and meeting all his colorful friends in Gibraltar. He was a character and a lot of fun and my whole mood towards the trip changed. Our captain was Terry Hewitt and the first officer was a Canadian pilot from Calgary in Alberta, Canada, 'Ollie' Oliver.

The flight to Malta was long and uneventful, and we opened the aircraft doors to warm air, sunshine and waving Palm trees. The crew taking over the airplane were waiting for us and we spent a few minutes talking to them before climbing aboard the waiting crew bus. They were working the airplane to Benghazi, Tripoli and Palermo which we were going to do after our two day layover in Valletta. On our short drive to the terminal for customs control we passed the airplane going to London. The crew waved at us as we passed.

It was G-ALWF.

I was in mid-sentence with Ollie Oliver when I saw it, and my conversation cut off like a knife. I must have turned quite pale and my silence caused him to look at me with a puzzled expression and say. "Are you OK –

what's wrong?"

"Yes, yes of course, I'm fine," I said somewhat embarrassed. "I am just happy to see that airplane going to London so that we do not have to fly it down here."

"It has been sold to Channel Airways in Southport, I understand," the captain chimed in. "That is why it is going back to London. The airplane we brought down will be staying here to replace it."

I wanted to hug and kiss him in thankful delight – but of course - I did neither!

Our two days in Valletta were long, lazy and hot. We lay out by the pool and toasted, basted to perfection with baby oil and iodine. The hotel was a delight – old spacious and very elegant.

The Island is bathed in history. I loved every minute of being there; dreaming of the Knights of St. John defending the Island from the Turkish invaders in 1565, while protecting the Crusaders on their way home from the Holy Wars. All this taking place while being under the protection of the King of Spain who ruled the Island in the 16th century. I soaked it all in.

That night at dinner in Valletta Charlie said, "We should go to the Blue Grotto tomorrow – make a day of it. Take our swimming trunks and a towel, swim and check out the caves. Then we could have lunch in Marsaxlokk and go through the Market. We can catch the bus down at the corner – it will only take about an hour to get there." Great idea – we were all on board with that idea and arranged to meet for breakfast at 8:00 ready to start our day of adventure.

It was a perfect day. The weather was hot, the water in the Grotto a perfect clear aqua and the caves a welcoming cool stillness. Lunch was delightful under a palm tree on the patio of a small cafe in the tiny village of Marsaxlokk. We ate conch salad with avocado and shrimp on Italian bread, and drank local beer and wine. Life does not get better than this – besides I was in love with Ollie Oliver.

The next morning we had a scenic breakfast at 7:00 on the hotel terrace overlooking the yacht harbor and the city of Valletta. Then we all piled into our crew bus for the ride to the airport. We were flying to Benghazi and on to Tripoli in Libya and staying the night. When I had the layover in Gibraltar with Charlie, we also went across to Algiers in Morocco which was a lot of fun, but this would be my first visit to the capital city of Libya. I was really looking forward to it - especially with such a great crew.

The flying that day was uneventful. The airplane was not 'sick' and performed beautifully. We had no turbulence and the passengers were mostly government people returning back to work after vacations or time off in Malta, Frankfurt, London, or even the USA.

Ollie landed the aircraft in Tripoli and I teased him later that I had to put my foot down outside the airplane to make sure we had landed – it was so very smooth.

The trip through customs and passport control upon

arrival was anything but smooth, however. We went through a 'Crew' line, but the rest of our passengers did not and it was loud, and fraught with trauma of every kind. If you want a peaceful transition you must produce greenback dollars 'under the table!' There was much shouting and waving of arms and running in every direction before people were allowed to slowly exit the sweltering terminal with their luggage. Raggedy children and stray dogs would appear from who knows where, to beg for food or some loose change.

In contrast, our hotel was small, quiet, clean, and right in the center of town. To get there we drove like lightening through crowded narrow streets. Some with houses built on bridges right over the top of them. This, apparently, was for oversight and defense of the ancient city when in time gone by enemies would attack from every side. It gave quite a claustrophobic feeling to the whole environment. Upon reaching the hotel we entered a small lobby, checked in, and agreed to meet at six o'clock to go to dinner.

We had two hours to clean up and relax before meeting again downstairs. I pulled off my uniform, took a long hot shower and then curled up on the bed to read up in my travel book, about Tripoli and this area of north Africa in general, for I had never been there to stay before. One of the things you find out pretty quick, should you decide to pick up a history book about this region of the world is that 'History' is a bottomless pit. Where do you start?

In the case of Tripoli, I saw that in the year of 551 an earthquake and tsunami completely destroyed the city!

Apparently it has always had a colorful past even before it fell to the Crusaders in 1109. It has a large very active port and has done much trade both East and West with countries in and around the Mediterranean. It also has always had a very busy Caravan trade route out of the city across the deserts and mountains towards Syria and beyond. It had made quite a name for itself by exporting textiles – cotton, silk, and especially velvet. Not only that, but apparently soap, in every possible variety, was made there and sent all around the known world.

What fun. I made a note to find someone to go to the Bazaar with me tomorrow. BEA had a strict rule that crew members were not to go to markets or bazaars alone on any layovers outside of Great Britain.

There is safety in numbers especially in locations like Tripoli. That goes for restaurants after dark too, apparently, because we did everything together as a crew, if we left the hotel. It felt good to me – I liked everyone on my crew – especially Ollie!

At six o'clock I put on a light blue summer dress and comfortable sandals and ran down to the lobby. The captain was asking the young man behind the counter if he would please arrange for a cab for us and also if he could recommend a restaurant that served good local food and wine. There upon followed much collaboration and merry discourse between him and another person in the office. It seems they could not make up their minds between them in spite of much laughter and hand waving.

Finally a decision was agreed upon and a horse and

buggy cab pulled up in front of the hotel and we all piled in. The young man from the counter explained to the cabbie in Arabic where we wanted to go and that he needed to wait for us at the restaurant for the ride home. All was agreed upon and understood.

It was a fascinating ride to the restaurant through this ancient city. It was a beautiful warm evening with a gentle wind blowing which was a relief from the oppressive heat of the day. As a result the city was alive with people and bicycles and carts and animals everywhere. Our driver negotiated this chaos with the learned ability of a lifetime. Sometimes yelling out – I am sure in very colorful Arabic – at some child or cart driver that did not clear his path quickly enough for his satisfaction.

Finally we arrived at a beautiful old almost Mosque like building set back off the road with a wide palm studded entryway. It looked like something out of an old silent movie about Lawrence of Arabia.

We got down and the cabbie pointed at the place where he would be waiting for us. The captain gestured to him that he understood.

With that, he turned and led us into an open courtyard and on through the doorway of the large domed building we had seen from the street. We had to stop momentarily to adjust our vision, because, coming in from the sunlit street the transformation was intense. It was dark and the odor was overwhelming. Our vision adjusting, we caught our breath and proceeded on in single file, the captain striding purposefully onward. Turning a corner we came to a large square chamber that was full of men in the long white robes of the region, reclining in every possible position, on chairs or the floor and smoking opium out of large bubble pipes. This was the odor that hit us at the door that we were not initially able to identify. I stopped dead in my tracks fascinated by the scene unfolding through the heavy haze in front of us. I stopped so suddenly that Charlie and Ollie both cannoned into me from behind - they also were absorbed by the amazing sight we had inadvertently stumbled upon. Suddenly I was horrified to see one of the men stagger to his feet and come swaying towards us through the gloom.

"Good evening – I vos expecting you an 'ave a table ready. You come please ja." he exclaimed in broken English with a distinct German accent. He led us down a dimly lit hallway to a double door that opened up into a large beautifully decorated restaurant and waved us to a table. Several other tables already had customers eating enormous portions of unfamiliar dishes and drinking wine.

We sat down and a waiter arrived at our side bowing deeply. We ordered drinks and while we were waiting for them the strangely haunting music that was playing stopped and suddenly the other customers started to clap. We all turned to see what was happening and a trio of musicians came in and set up their instruments on the floor to one side of an open area, and started to play. Then a beautiful young woman came dancing into the open space. She was dressed in long silk pantaloons with lots of bells and jewelry draped all over her body and long black hair. She looked like she had just stepped out of a scene from *The Count of Monte Christo*. This was a belly dancer of amazing talent, come to entertain us – or should I say 'them' – I was the only female in the place!

Dinner was amazing – unrecognizable, varied, and delicious although I had no idea what I was eating. We all had a blast and ended up getting belly dancing lessons from the locals after much wine was consumed by all around the restaurant. The patrons apparently were delighted to be entertaining us – us from another dimension - and me the only girl!

We left the restaurant in wonderful spirits bidding our new friends a reluctant good bye, but anxious to head home. We retraced our steps down the long hallway – the odor of opium and the resulting haze getting ever thicker as we closed in on the den. Finally, we turned the corner and there they all were – but now it was three hours later and these men were 'flying high.' As we walked hastily by, many of the men started to get up and leering grossly, stumbled towards me. Ollie grabbed my arm and almost carried me at an alarming rate past the opening, and out the door. That was scary and I thanked him profusely as I tried to regain my composure.

The captain waved and the horse drawn cab materialized out of the night. We all climbed gratefully up and onto it. We had a lot of fun, but the stress and the wine were having their obvious effect. We wanted a quick uneventful drive home through now slowly emptying streets. It was dark. The narrow streets had no light except in the passing homes.

Suddenly the cabbie pulled the horse over and

stopped. He swung his legs over the seat so that he was now facing us.

"Captain, we need to talk," he said in almost perfect English.

The captain had just removed his boot and was pouring sand out of it onto the floor of the cab. He hesitated with the boot in midair and said, "Excuse me. Excuse me, you speak English?"

"Yes sir, I worked for the British army in World War II – I speak English." There was silence in the body of the cab. The horse stamped his foot wondering what was going on. He was not alone, but we were about to find out.

"I have been commissioned to ask you by several people here today. HOW MUCH FOR THE GIRL?"

The captain shoved his foot down deep into his boot, sat back against the cushioned back of the benched seat and looked the driver in the eye and said quietly but very firmly. "She is not for sale."

"Captain, be reasonable – every woman is for sale! This one VERY desirable – very fair skinned and hair like ripe, sun drenched oats. Also good long legs – I am able to do a very good deal for you."

"Now ye listen here laddie – ye are getting me Scottish blood up way too high fer ye'r own good. I have a desperate need for this young lassie to be with me tomorrow and I am telling ye I cannot – nay WILL NOT be selling her to ye or anyone of yer brethren any day soon. Take us back to the hotel this minute or I will be removing yer self from that seat, makin' Haggis out o ye and depositing ye at the side of the road amongst the beggin' trash."

The cabbie seeing that my captain was deathly serious, turned slowly around on his seat and with defeated sadness, clicked the horse into motion.

Tonight the Casbah did not get lucky.

Tripoli 1963

CHAPTER 19 -

FIXING THE UNFIXABLE

Ollie and I became a serious 'item' upon our return to London. We bid to fly every trip together after our return from Tripoli. Soon thereafter we moved into a small apartment together in Hounslow, close to London Airport. I was over the moon – now I knew exactly what I did not know with David. THIS was love and I was in love with Ollie Oliver and he was in love with me. We were inseparable – joined at the hip - at work or at home. Fred and Clifford loved him and my friends found him fun to be around. He was an amazing man and an accomplished pilot fifteen years my senior.

There was just one fly in the ointment – he was married and the father of three little boys, ages 7, 5 and 3.

When he left Canada he and his wife were separated, but both came from good Catholic families and divorce was not an option. When he took the job with BEA in London he was worried about not being able to see his boys, so he made his wife an offer. Come to Europe with the boys and I will set you up in a villa in Spain for the duration of my assignment. She jumped at the deal and so they were living in a beautiful villa with a cook and a nanny in Malaga, Spain. He would go down once a month to see the boys and pay the bills.

Our life in London was quite sparse – we did not have a shilling to spare but we were happy. We were so very happy. As time went by, one year – then two, my friends – and of course Fred – would wait – expect an announcement, a plan – a shoe to drop! There was no shoe to drop – there would never be a shoe dropping session! But they did not know our secret.

I was in agony. Ollie and I could never have a true life together. THAT WAS REALITY. When he was gone I would think about those darling little boys and I knew how much it hurt him to be separated from them. I did not know what to do. I loved him and I could not give him up. I did not have the emotional strength to move out and stay in the same universe with him - risk seeing him on flights or on layovers. I did not know how to fix this but I also knew – I must.

Enter Pan American World Airways.

Pan Am was desperate for people that spoke languages. They were a round-the-world airline and needed to be able to communicate with their passengers at 30,000 feet over anywhere in this world. To do that they needed stewardesses that spoke languages because Americans did not. America is a vast beautiful country in which everyone spoke the same language.

They came to London airport begging for help.

Ollie and I were returning from a trip when I picked up a flier from under the windshield wiper of my Messerschmidt in the crew parking lot. Ollie did not see it because he was loading our bags in the back, and I hastily shoved it into the pocket of my coat. When we were home and he had gone out to pick up some fish and chips for our dinner – I pulled it out of my coat pocket and stared at it for a long time. When I heard him returning I shoved the flier into the back of my underwear drawer.

It was several days before it once again saw the light of day. Ollie was flying an early trip to Glasgow and I was not to join the crew until later that afternoon. I took a long, hot bath and propped the flier up against the wall at the end of the tub. I soaked in that hot water for an hour or more and stared at it.

Could this be my 'fix'?

I applied to Pan American and waited anxiously for a reply, not really expecting to get one. They were so fussy. It was a well-known fact that they were very difficult to get into. They were 'The World's Most Experienced Airline' – America's Flagship carrier – their logo was as well-known as Coca Cola around the world. They would never hire a person like me. What could I possibly offer a company such as that?

I came home from a Berlin trip. BEA had been utilizing me to fly the Internal German Routes again. Ollie was in Malaga. I picked up the mail and found a letter with the Pan Am logo on it. It was addressed to me and came from New York. I set it aside on the dresser – checked the rest of the mail – mostly bills, and went in and took a bath. It was many hours later, when I came back into the bedroom, that I saw it sitting there against the mirror. I slit it open with trembling fingers, fully expecting it to say "Miss Wheeler, thank you so much for your interest – but we, have no interest!"

How many times has it happened to me now in my short life, that when fate slammed a door shut the good Lord opens a window? Perhaps Pan American World Airways was that window.

The letter invited me to come for an interview the following Friday at 10:00 am at the Athenian Court Hotel in Green Park near Buckingham Palace in the heart of London. I could not believe it. I read the letter again. It said the same thing.

I quickly ran to my overnight bag and pulled out my schedule. I was returning to London from Frankfurt at 0800 that morning. It would be tight and I would not have time to go home and change – I would have to go to that interview in uniform. In a way it was convenient because Ollie would just think my flight had been delayed when I was late getting home. I had not told him that I applied to Pan Am because I really expected it to be a moot point, besides I did not know how he would take it. Neither one of us had the heart or the courage to leave the other. This would put the whole Atlantic Ocean between us. I had to find the strength to go through with it if I could.

My flight from Frankfurt got into London at 0745 – a little early. I ran to the bathroom and freshened up and then out to find Putzi. I drove into London as if in a dream. What was I doing – how was I going to explain this decision to Ollie? It all did not matter because they would not hire me anyway. I parked Putzi in the back of the hotel in the delivery yard and made my way into the lobby. The desk clerk pointed me to the downstairs ballroom where the applicants were to meet. Sure enough there was a large Pan American World Airways sign with their blue ball logo sitting to the left of a large open door. I went in and introduced myself to a lady in Pan Am uniform and showed her my letter.

"Good Morning Gisela. My name is Amy Peroz – are you coming or going on a flight?"

"Please forgive me for coming in uniform, but my flight got in this morning and I did not have a chance to go home and change. I did not want to risk being late." I stammered as I took her outstretched hand. "No problem at all Gisela, and as a result you are here with time to spare, which is great. Please step onto this little scale for me, and then you can sit over there by the window and fill out these two forms. When you are done you can bring them back to me and I will assign you an interviewer."

I did as I was told and when she had made a note of

my weight I went over to the large bay window facing Piccadilly, lowered myself into a cozy easy chair and started to fill out the forms Amy had given me. When I finished I got up and took them back over to her and waited patiently while she dealt with several other applicants arriving at the door. The large ballroom was starting to fill up. There must have been fifty or more young ladies there when I arrived and now we had almost doubled that.

Amy took my paperwork from me and smiled. "Gisela, please take this folder and go straight down the hallway to room 325 – and Good Luck."

I was floored, excited and scared to death. I had expected to wait for some time in the ballroom before being called. I floated down the hall and knocked on the half open door of room 325. A pleasant male voice said, "Come in. Hello. You must be Miss Wheeler. My name is Joe Hale – please sit down and you can tell me all about yourself and what you are doing in that uniform." He closed the door and took a seat across from me behind a small elegant desk.

Joe Hale was a delight and very easy to talk to. He was a steward with Pan Am based in Miami and he soon had me laughing as we exchanged in-flight stories. We talked for quite a long time. Until the phone rang – he excused himself and answered it. Then he turned to me and said, "Gisela, I am so sorry, I have to cut this interview short because we have so many people to see but I loved talking to you. This has been a great pleasure and I plan to recommend that we hire you as a stewardess and exchange that dark blue uniform for a light blue Pan

Am one." I jumped to my feet like a scalded cat and held out my hand to him. We were having such a fun talk – I did not realize that I was being interviewed! "Thank you, thank you so much." I stuttered awkwardly turning to go.

"Gisela I must ask you to go down the hallway to the left when you leave me. Then take the last door to the right – it will lead you back into the lobby. We don't want you to go past the other applicants waiting to be interviewed because you are going on to the next interview phase. We are going to see about five hundred people today, but only about a half dozen will make it to Phase II. Congratulations and I hope to see you down line. We will call you when we have a date and time." he smiled and shook my hand warmly and I floated out of the door.

Ollie arrived home later that day with an arm full of tulips and daffodils. He had come back to London via Amsterdam and the tulip fields were in their full splendor. He always came home with some flowers or a fun trinket for me from where ever he had been. He is so sweet and I loved him so very much. How pray tell was I going to be able to go through with this crazy idea.

We both had two days off and we had an invitation to go and visit Fred and Clifford in their new home in Wimbledon. I would have to find the right time to tell him – but it could not be now.

The letter for the second interview came the following week with a date and time. This was a slightly different format although in the same hotel. I would be interviewed by a panel of five individuals.

I checked our schedules with a racing heartbeat. Ollie would be in Malaga and I had a Paris turn-around. I could call crew scheduling and ask them to please take me off that trip and give me an airport stand-by in the afternoon, either that or they could schedule me for a trip the following day which I had off.

The big day dawned bright and clear and I set off in terrified good spirits. It was all very exciting, but thoroughly useless because they would never hire me. This was THE GREAT PAN AM after all. I had completely convinced myself of this and as a result had not told anyone about what I was doing – not even my Mother and Clifford let alone the man in my life.

I checked in to the Pan Am reception desk and took my place with three other girls. We all silently smiled and secretly eyed each other, trying to get a hint about our competition. I started filling in my bids for next month's flying while I waited, not wanting to sit there in that awkward stillness. As a result I was so deeply engrossed in my task that I failed to react when my name was called.

"Miss Wheeler. MISS WHEELER - this way please." I jumped up so quickly that I spilled my entire purse at the startled lady's feet.

Not a good start.

Pull yourself together.

You can do this.

I followed the receptionist into a large room with five people seated in a row facing a single empty chair. The room was bright and cheerful with large Pan Am posters from all around the world hanging on the walls. I was ushered to the empty chair and invited to introduce myself. Then my interviewers all introduced themselves in turn. The last being a strange stern looking little man in a dark suit and black tie. I did not think that was a good omen and he, unlike the other four was not at all friendly or welcoming. Also, he was the only English person of the group – one was German, one was French and the other two were American.

They all asked me questions in turn. About my life – what I had been doing and why did I want to join Pan Am. Did I not like BEA and so on. I explained that I liked my job very much, but the flying was restrictive. Because of my German background I was flying German domestic routes a lot and also BEA's routes were limited to Europe and North Africa. I was interested in flying to Honolulu. It was my dearest wish to fly to Tokyo and Hong Kong and Bangkok all of which was not possible with BEA. They were very friendly and we talked about the posters around the room – some of the destinations I was very familiar with and they liked that. Some of the conversation had switched to German which I only realized when the only person that had stayed silent and said not a word was Mr. "Doomsday" Bamberger! He did not speak German and finally come to life and asked sternly in English.

"Miss Wheeler. Where did you go to school – and did you graduate?" It came kind of out of left field and broke the pleasant lighthearted banter of our previous conversations

"Yes sir. I graduated with eight subjects for my Certificate of Education at 'O' level. I went to James Allen's Girls' School in Dulwich and elected not to go to Oxford as planned, but to follow my heart and join an airline. And that is what I did, and why I am here now."

Mr. "Doomsday" Bamberger shuffled his papers around and I distinctly saw when he made a check mark on one of them.

"Thank you Miss Wheeler – that is all for now. We will send a letter and let you know the outcome of this meeting."

And with that it was over and I walked out of the hotel into the noise and bustle of a busy Piccadilly.

CHAPTER 20 -

TRANSITIONS

It was exactly one week later. I was sitting in my cozy chair by the fire listening to the radio. *Mrs. Dale's Diary* had just come on at four o'clock – it was four ten. I had the day off and was kicking back and totally relaxed, enjoying this stupid program I had grown up listening to with my Mum. I heard Ollie arrive home from a Dublin, Edinburgh, London trip. He had stopped to pick up the mail, shed his shoes at the door (I had never figured out why he did that) came in and kissed me Hello.

I switched off *Mrs. Dale*, jumped up and hugged him. He had only left at six o'clock that morning, but I was always so happy, and if truth be known, relieved to see him home again and safe. This was really silly, because he was an excellent pilot and air travel was a lot safer than driving your car – BUT since Malta I always had some slight unease. Not for me when I was on a trip, but when he was on a trip without me I was stressed till he came home.

We hugged again and he said, "Weather was a real pain today – bumpy, bumpy, bumpy all the way home! Did you have fun doing nothing all day – how is Mrs. Dale?!" He was being silly, of course, but I loved his teasing. "I picked up the mail – there is something for you from Pan Am – what on earth could they be writing to you about?" He gestured at the mail he had thrown on the table, turned and left the room tearing off his shirt and tie to take a shower. Funny. That was always the first necessity after getting home from a trip. Did not matter where we had been or how long we had been gone. We had to get under water, and it cost us many extra English pounds on top of our rent. I smiled at the thought.

When I heard the water come on in the bathroom I squeamishly pulled the letter out from the bottom of the pile. I held it in my hand for a long time – just savoring the idea that THE GREAT PAN AM was sending me a letter. The contents of which were of no consequence for I did not expect anything of interest or value to be enclosed in it. There is no way it could be holding good news. Not for me.

Miss G.I.M. Wheeler,
27, Summerhouse Avenue,
Heston,
Middlesex.

October 15th 1963.

Dear Miss Wheeler,

We are happy to inform you that you have been selected for employment as a Stewardess with the Overseas Division of Pan American World Airways. Actual employment is subject to the following:-

(a) Your satisfactorily passing our Company medical examination.
(b) Your obtaining a U.S. Immigration Visa in time for the training class to which you will be assigned.
(c) Our obtaining a satisfactory reference.

In connection with the Company medical examination you are requested to telephone our Company doctor, Dr. R. Wambeek on SLOane 8045, to arrange an appointment. This examination will take place at his surgery, 48 Sloane Square, London, S.W.1. The nearest tube station is Sloane Square and the surgery is situated next to the Royal Court Theatre. You should note that any expenses incurred in attending this medical examination are for your own account, except that the cost of the actual examination is borne by this Company.

As we plan to assign you to a training class early in 1964 you should commence immediately your application for your non-preference Immigration Visa for the United States. In this connection, we are enclosing a visa application, which you should complete and forward with the two copies of our letter of support, to the nearest American Embassy or Consulate to your home.

Prior to your departure for your training class you should also have your occupation shown in your passport changed to read that of Stewardess.

At the present time, you should not take any action to terminate your employment. This should only be done when you have been advised that you have satisfactorily passed our Company medical examination, have an indication when your visa will be available and have been assigned to a definite training class. At this time we shall not contact your present employer for a reference.

/Cont'd. . . .

OVERSEAS DIVISION · LONDON AIRPORT NORTH · HOUNSLOW · MIDDLESEX · ENGLAND · REGENT 8474

Acceptance Letter from Pan American

It was about thirty minutes later that Ollie came back into the room after his shower, wrapped in a cozy bathrobe and carrying two glasses of wine. What he found when coming through the door was in no way what he had expected. I was sitting at the table with unopened mail all around me with my head on my arms crying as if there had been a death in the family. In a way that is absolutely what it was. Ollie put the drinks on the sideboard and rushed over to me, pulling me up to him and wrapping his arms tightly around my shaking shoulders.

"Gi, what – what in the world is it. What has happened – has somebody died?" I tried to pull myself together but just shook my head – YES!

"What – who?" he said very quietly releasing his hold on me.

"US!" I said with tears streaming down my face and waved at an open letter lying face down on the table. He reached over and picked it up – very slowly, turned it over and began reading. The color drained from his face and his hand shook very slightly. He looked up and our eyes met – held – for a long moment. The silence was heavy and very intense.

My heart was breaking and when I saw the pain in his face and eyes – I knew his was too.

Five days later I got a telephone call from Pan Am giving me a date and time for my visa interview and medical with the US Embassy in London. Nothing could proceed if I failed either one of these.

After the heartbreak of the letter informing me of my acceptance to join Pan American World Airways as a stewardess – actually the heartbreak of having to explain it all to Ollie – I had gone to stay with Fred and Clifford. I needed some space and I needed some alone time with my Mother to regain my equilibrium and focus. They too, of course, where in shock, since I had not told them or anyone else what I was doing. My Mother tried very hard to be positive – because she knew very well why I was doing it. If I had applied to Lufthansa, as planned, or even BOAC, the result would have been the same – I would constantly be running into Ollie and she knew that would never work. I had to put nothing less than the Atlantic Ocean between us. How very sad.

On the agreed upon date and time, Fred accompanied me to the US Embassy in Grosvenor Square. It was a very grand, if intimidating place, and I was happy to have her come along. She helped me fill out a stack full of paper work regarding my German background and subsequent English transition. When we were done I relinquished both my British and German passports and my swastika covered birth certificate, to a young man behind a very large antique desk. He gravely looked everything over and then waved me to a window seat to await being called for my medical. I explained to him that I would need at least one passport back before leaving because I could not work without it and I had a Barcelona flight the next day. He assured me that could be arranged and was true to his word when he handed me back my British passport after my medical.

It all took most of the day, and when we arrived

home tired but relieved it was over, Clifford suggested we go out for dinner and celebrate. This was an idea that so far had eluded me in my pain. I had been so engulfed in sadness I had not considered that I literally had the world at me feet. Celebration was overdue.

When I got back from Barcelona, Ollie went to Malaga and while he was gone, I moved out. If I stayed with him my resolve and courage would crumble and I would not be able to carry this through. Although devastated and hurt, he understood why I felt I had to do this, and he did everything he could to help.

It was now early December, Christmas loomed, and the wheels continued to turn slowly. I was assigned a class date of February 3rd 1964 pending my passing my medical and acquiring a visa. Several days later I got good news and bad news.

My visa had been approved but could not be awarded because I had failed my medical. I could not believe it. I was the most healthy of human beings. Why in this world could I have failed the medical? I called the number offered on the letter and got through to the Medical Department of Pan Am. Yes, they were very sorry to inform me that even though I was very healthy and had never called in sick to BEA, I had very unhealthy tonsils, and they would have to be removed before I could pass and get my visa!

The world momentarily stopped spinning.

December 17th, 1963

Miss G. Wheeler,
27 Summerhouse Avenue,
Heston, Middlesex.

Dear Miss Wheeler,

Further to our recent telephone conversation we have pleasure in confirming with you that, subject to your obtaining your non-preference Immigration Visa for the United States and satisfactorily passing our Company medical examination, you have been assigned to a training class in New York on:-

FEBRUARY 3RD, 1964

We would further advise you that we plan on having you depart for the United States approximately two days prior to the commencement of your class. Actual travel details will be forwarded to you nearer to your departure.

Yours sincerely,

E. Bamberger
Personnel Representative

/PR

OVERSEAS DIVISION · LONDON AIRPORT NORTH · HOUNSLOW · MIDDLESEX · ENGLAND · REGENT 8474

Class Date Letter from Pan American

Two days later I sat in the waiting room of my doctor's office in Dulwich Village. Dr. Harvey had been my doctor ever since my arrival in England, and had dealt with every childhood illness known to man including my recurring tonsillitis. I explained my predicament to him and he reached behind his desk and lifted a very large file folder into the space between us.

He turned to the very last page and said, "Well alright, Gisela. I can do it on the 10th November 1965."

I looked at him and actually laughed out loud in my confusion, as that would be two years from now. "I don't understand doctor – I would need you to schedule me between now and the New Year – I have a class date in early February."

"Don't be silly Gisela that is not in the least bit realistic. It is not an emergency so you would have to go to the end of the line."

"Forgive me doctor, but IT IS AN EMERGENCY – could you do it if I was to be a Private Patient?"

He looked at me sternly for a moment and then went to his filing cabinet and pulled out a folder with three pieces of paper in it. Turned the last piece over and said, "I can do it for you at two thirty next Thursday afternoon at Kings College Hospital. The only thing is, if you are serious about this – you would have to have your wisdom teeth out before I could do surgery on your tonsils. Those teeth are breaking through your gums and could cause a terrible infection."

I left the doctors' office in a dull haze. I do not remember the drive home to Wimbledon. I felt like I had stepped off a curb and fallen head first into a pit. Fred listened and greeted the news in stunned silence.

"The world has gone mad, I think," she stated unknowingly reverting back to our native German. "Go and call Anne. Her father will help us out I'm sure when he hears what is happening." Anne had been my dearest, closest friend throughout my JAGS school days. She lived on the corner of the Village and Lordship Lane and her father had been my dentist for as many years as we had been friends.

I went to see him the next day. Her Dad was the sweetest man and he was very happy to see me – we had not seen each other for almost two years, or since I had moved permanently closer to the airport. He was very excited at my news and said of course he would help.

"Open up Gisela, let's take a look at the problem." I did as bid and he took a long hard look in my mouth. Then he went over and checked a calendar and his file. Three days later, on Monday, I found myself back in his office to remove my wisdom teeth.

Fred had come to drive me there and back and visit with Anne's mother. Anne was there to hold my hand and ease the trauma with a little brandy after the 'deed' was done, while everyone else was enjoying tea and lively conversation. Anne's mother and Fred were saying how much they missed the days when we all lived close to each other. Anne and I would come rushing in to each other's houses after school and the families would meet at the 'Dog' - The Crown & Greyhound Pub - in the Village for beer and sausages on a Saturday afternoon.

I had my teeth out on Monday, that was four days before Christmas Eve. Thursday was the day BEFORE Christmas Eve. That was the day I checked into King's College Hospital to have my tonsils out!

I awoke from the surgery to see my Mother sitting in a chair reading. She looked up when she saw me stir and came over to the bedside.

"Wie geht es Spätzchen?" she said quietly and put her hand on mine and squeezed gently.

I rolled my eyes and made a face – I could not speak. It would be eight days before I would be able to leave the hospital. I thought of the days when kids would have their tonsils out on the kitchen table and the next day they would be fine. Not so if you are an adult. One of the problems is that you cannot – or do not want to eat. The hospital tried to tempt me with Cornflakes. I found the idea to be quite amazing since I had open wounds all over my mouth and throat. I begged my Mother to bring me something 'soft' that would go down easier. She came with a banana. I wanted to hug her – peeled it and took a bite, I was so very hungry.

I almost lost my life with that bite! It stuck in my throat and I could not breathe. Eventually, back came the Cornflakes and low and behold, they went down like a charm!

We celebrated New Year's quietly that year with just Fred and Clifford, Dick and Betty, and me at home in Wimbledon. It was very nice and New Year's Day was my Mother's birthday. It was a Sunday that year which made her happy – she was used to it being a holiday in Germany and could not get over the fact that it was not a holiday in England.

On January 2nd I called Pan Am and informed them that I had my tonsils removed and could I please make another appointment with their medical department in order to get my visa approved. They fit me in at noon the following day at their London airport medical office.

Three days later I got my letter of acceptance written the day of my visit.

I was stunned. Suddenly with this letter reality sank in. It would mean leaving everything and everyone I knew and loved. Move by myself, at twenty three years old, five thousand plus miles away to another world. The mere thought of it took my breath away.

Could I do it?

Would I do it?

Transitions

CHAPTER 21 -

FLYING ON WINGS OF MEMORIES

It was with a heavy heart that I drove to the airport on a dark, rainy winter day the second week of January 1964, to hand in my resignation to my supervisor at BEA. I pulled in to the crew parking lot and came to a screeching halt while turning a corner looking for a vacant spot. A captain I knew well leaned out of the window of his new Humber Hawk – and while blowing his horn yelled at me with much good humor.

"Damn it Gi – paint yellow spots and a Union Jack on that bloody thing. I almost shot you down!" I waved merrily at him as I drove by. He had been a fighter pilot in the war and took great glee as did many other crew members, in teasing me about my Messerschmidt. He managed to lighten my mood for the task before me and I entered the crew lounge with a more positive, upbeat state of mind. To say my supervisor was unhappy to hear of my plans, was an understatement to say the least. BEA was still short of stewardesses – especially ones that were language qualified and I had been such a help with the Berlin flying. It made me sad that I would now make the job of covering that flying and training program very difficult. I left with a heavy heart and hurried back out to Putzi before I saw anyone else I knew. I did not want to talk about my decision and plans – I just needed to get it done before I changed my mind.

When I returned to Wimbledon Freddie told me Anne had called – she wanted to get some friends together for a Good Bye, Good Luck Party at the 'Dog' for next week. Please phone her back. I did and we had a good time planning it. I had some airport friends I wanted to ask – two agents that I had worked with from Alitalia Italian Airlines that had an office next to Lufthansa, and of course my Lufthansa buddies. I had stayed close to all of them even after I started flying for BEA. And of course, my old roommate Sheila, who had been directly responsible for that.

The party was wonderful and very emotional. Everyone was supportive and happy for me with a tinge of trepidation. It was such a ginormous up routing, scary decision. Mainly because of the distance involved, but some were even very envious of the opportunity to fly for Pan Am, and said so.

It was close to one in the morning when we started to

break up. Sheila did not have a car and I had picked her up to come to the party, so we got back into Putzi for the long ride back to Hounslow. We were in very good spirits. The traffic was light and we chatted merrily – flying along on wings of memories, reliving fun times that we had spent together and would remember, always.

Suddenly a cyclist with no lights on his bike, came shooting out of a side street. I have no idea how I managed to avoid hitting him. We were approaching a corner, a long bend to the right, and as a result when I tried to correct our momentum I over corrected and yanked the steering over to the left. The back wheel slid when I applied the brakes, hit the curb and we started to roll. When the canopy of the Messerschmidt hit the road it shattered, splintering and bursting into a gazillion pieces - like someone had stuck a balloon with a pin. We rolled twice, maybe three times and amazingly ended up upright. I looked around – dazed and shaken. I turned around to check on Sheila. She like me, was sitting there dazed with a cut on her head that started to bleed.

To my utter amazement an ambulance was coming towards us and several people had appeared as if by magic out of the night. We were helped out of the car by many willing hands and loaded into the ambulance. The door was shut and we drove for three minutes. Then the door was flung open and we were being transported into an emergency room. It turns out I had managed to roll that Messerschmidt into the front entrance of West Middlesex Hospital! In the hands of God – again!!

I had a broken wrist and a very badly cut and bruised knee which had caught the steering column during the rolls. Sheila – thank heaven, only had that cut on her head. She had managed to duck down and hold on to my seat and it had spared her more injury. It was a miracle we were both alright and in good hands.

Fred and Clifford came and picked us up at six in the morning. We drove Sheila home and then drove on to Wimbledon leaving the wreck of my precious Putzi behind.

Now my predicament was not just emotionally painful, it was physically painful too. I spent most of the next day just vegetating in bed. At lunch time Fred came into my room with a tray. Marmite and cucumber sandwiches – my favorite, and of course, tea. It was a very special Mother daughter moment. She propped me up with pillows so that my wrist was comfortable, sat down in the chair across from the bed and said in German.

"Eat, and then let's talk this through." I did as I was told while she poured the tea. "Are you one hundred percent sure this is what you want to do?" she said after settling back down in her chair across from me. There were so many emotions going through my brain, it was hard to concentrate on just one – but THE most important one of the moment was what she wanted to discuss.

"Ja Mutti, I must go. Ollie and I have agonized over this for over a year. There is no fix. I cannot ask him to cut himself off from his boys, and that is what it would take. I know that I will never find another love like the one we shared, and neither of us has the strength or the will to leave the other. We could not bear the risk tripping over each other in a crew room, down line, or heaven forbid on a ten day Malta trip. We could even be scheduled to fly for a whole month together, in Berlin. We cannot do it. It could not happen. This is the only fix."

The tears that I had denied myself for so long started to flow. It was because I had been so stoic about my decision that my Mother needed to have this conversation with me, and was relieved to see this reaction. It was such an enormous decision and she needed to be sure that I had really thought it through and could go on with it without regret. She kissed me on the forehead picked up the now empty tray and left me alone to my thoughts and emotions.

There were many. I went back a long way laying there in the semi-darkness of a late winter afternoon. I had so much to be thankful and happy about. Lady Luck had waved her magic wand over me so many times since my birth. The nightmares - those tricks of memory that often catapulted me back to my childhood in Hitler's Germany, slowly faded with time in my new homeland. The dark traumatic days of fear, death and survival that we had endured during that terrible time so long ago, no longer haunted me in my dreams.

I had lived and experienced a charmed youth. Growing up in a wonderful, loving new family, that awarded me the ability through love, encouragement and education to reach for the stars and achieve anything my heart desired. I had done so.

My heart was full of gratitude and love, for all that

had past and all the wonderful people that had helped get me to this moment in time.

It was now time for The Eagle to be strong and without fear, spread her wings and fly over the ocean to another new beginning.

Flying High

Book III of
The Eagle Must Fly
An Autobiographical Trilogy

Pan American World Airways – 1964 Graduation

PREFACE

Time is once again suspended.

I look back on my childhood and ponder – how many times have I been blessed with love and opportunities I could never have imagined. In my childhood in Hitler's Germany, had my Father lived and raised me, what would my life have been like. Would I have enjoyed any of the opportunities thrown my way the way they were as a result of me growing up British. I think not.

I am who I am and all the things I have accomplished because my Father gave his life to the Third Reich, and Hitler, to be cannon fodder in the hell of a Russian winter. All because he was unfortunate enough to have tuberculosis. The same Hitler and Third Reich that failed its people and forever trashed them in the eyes of the world. This failure would bless me by being found and rescued by a British soldier who fell in love with my Mother in the shell and rubble of what was left of a city.

Hamburg, Germany was a Phoenix that would eventually arise and become reborn in the tender hands of the Allies and The British Control Commission.

British Army Captain Arthur Wheeler spoke fluent German and as a result was commissioned and sent to Germany to help the city of Hamburg and it's people. He did an amazing job for two years helping to rebuild that war torn city. Subsequently he rescued my Mother and me by taking us to live in a new country and giving us a new beginning and me a rebirth.

As a result I grew up British, and thanks to Arthur Wheeler and the life and education he provided for me I was now able to follow my dream. To fly.

The Eagle will fly.

Oh Yes – The Eagle WILL fly high.

CHAPTER 1 -

NEW BEGINNINGS

Our dramatic arrival into the beautiful city of New York was anything but beautiful. The airplane hit the runway with an enormous THUMP. It slid and bounced and groaned till it finally came to rest at the insistence of the cockpit crew, and not of the incredible downpour of water and wind engulfing it. Our approach to the city had been decidedly 'crab like' as we fort the heavy crosswind from the Atlantic Ocean. The long, time consuming taxi into the gate, felt as though we were rolling all the way back to London at zero feet. The weather sadly obliterating every chance of a view of our new home. The north eastern sea board of the United States is not the most desirable location in the throes of early February. The weather was mean, cold, wet and unwelcoming – but we had arrived. We were here.

New York, New York.

We were met after our release from customs and passport control by a member of the training team from the New York crew base who politely piled us into a crew van, and without fanfare, drove us through blinding rain to our new home in Kew Gardens, Queens, New York. The building housing the Pan Am trainees was just one of a gazillion in that concrete jungle. We were in awe of our surroundings as we literally blew out of the crew bus and strained against wind and rain to retrieve our bags.

Once inside the welcoming lobby of 'The Forrest Park Towne House,' we were assigned our rooms. Chloe and I were to share a two bedroom apartment with two other trainees already there. The other four trainees from Sweden were given another two bedroom apartment on the floor above ours. Then, with instructions to be down in the lobby at seven o'clock sharp in the morning, for pick up to the training school, our happy Pan Am greeter left us with a cheery wave to find our way to our assigned apartments.

The apartment in The Forrest Park Towne House was quite luxurious to our naive European eyes. It was fully furnished including linens, dishes and cooking utensils and each apartment accommodated four trainees. It was very comfortable and also had a resident supervisor living on the premises in case we needed help or assistance with anything. The facility was specially selected by Pan American to house its trainees for the four week training period because it was very convenient to the airport and transportation facilities. We had been informed in our 'acceptance letter' that we would be required to reside there for the duration of training – even if the trainee was local, and from the New York city area.

We arrived at our designated apartment to find the door ajar and music playing. Chloe pushed the door open with her suitcase and I pushed my head inside and called out "Hello – anyone home?" A young Asian girl came running from somewhere inside, grabbed the door and flung it wide.

"Yes, yes. Hello – are you our new roommates? Oh welcome, welcome. Come in please – we have been expecting you, but the weather has been so bad we were afraid you could not land today. Micky, Micky our new room mates are here. Micky is in the shower," she explained excitedly helping me in with my bag. "My name is Akita – I am from Tokyo and Micky is from Manilla. Come in and let me show you your room." With that she grabbed a bag and pulled it behind her into a half open door.

Our bedroom consisted of two twin beds, two bedside tables with a chair beside each and a large oak dresser. A door opened into a walk in closet – the likes of which neither Chloe or I had ever seen, and a window looked out onto the street and the labyrinth of tall buildings that made up Queens, New York. It looked clean and welcoming. We were home.

The next morning, at six forty-five AM, thirty-five young ladies from all over the world descended into the spacious lobby of The Forrest Park Towne House. The excitement was intense. Everyone introducing themselves to everyone close by, and exchanging information about past present and future hopes and plans. All the trainees had passed mandatory checks and qualifications two pages long, including height, weight, languages etc., before the interview process in their homelands had even begun. This group of young ladies was without a doubt – the crème de la crème of the world. A large bus was parked and ready out front, to transport this treasure to the Pan American Training School at Idlewild International Airport – the World was at our feet and waiting.

When the door of the elevator opened onto the lobby and Micky, Akita, Chloe and I spilled out into the mayhem, we were overcome and unprepared to see so many strikingly attractive, glamorous girls in beautiful clothes looking like they had just stepped out of a magazine. Even more amazing was the multitude of languages being spoken casually and with ease. French, German. Italian, Portuguese, Russian, Polish, Arabic and Hindi – even Japanese and some I could not identify. It was very exciting and exhilarating to join and find ourselves part of this diverse group.

On the dot of seven a gentleman entered the lobby and the noise level closed down slowly, as though someone was turning off a tap of running water.

"Good Morning Ladies. My name is Bill Viola and I am delighted to be your instructor for the next four weeks. Please be good enough to board the bus and we will be on our way to the training center, and welcome on board. We are delighted to see you and cannot wait to begin."

Sitting on the bus on our way to training, Chloe and I were able to look around and check out our fellow classmates. The thing that struck us immediately was – we all looked alike! In spite of the fact that we came from amazingly different backgrounds and parts of the world – we all looked alike! We had been cloned. Our hair and make-up almost identical – that was the Pan Am way.

The training center was very close to our residence and would only be a short bus ride away, which was the idea. On this first morning we were all picked up however, to make sure we all got to the training center at the same time to acquire all the necessary information regarding training. Sitting there and listening to all the happy chatter I was overwhelmed by the amount of talent on this bus. Not only were the languages from all over the world, but many of these ladies were nurses also. It was exhilarating to know that the passengers on our long around the world flights, would be so very well cared for, no matter what the situation. A child's ears hurting, a heart attack, a baby born, or even – heaven forbid – an emergency.

We were on our way to flying high.

Book III: New Beginnings

ABOUT THE AUTHOR

Gisela Wheeler Scofield was born in 1941 in the beautiful city of Hamburg during some of the darkest days of our history – Hitler's Germany in World War II. She rose like a Phoenix out of those ashes, as did the city of her birth to fly the world with Pan American Airways.

Her destiny let her to grow up in England as the casualty yet survivor of the war, and to go on the emigrate to America as a Stewardess when Pan Am came calling. The world was her oyster as she flew serving royalty and the famous. She also learned to deal with sadness as she met and flew young men into battle in Vietnam. Then, the exhilaration and joy when she flew them home.

She married an American Naval officer she met in Bangkok and had two boys. They then adopted a baby daughter from Korea. Gisela has eight grandchildren and lives in Texas with her two miniature dachshunds Sandi and Bindi.

Made in the USA
San Bernardino, CA
10 May 2018